SCIENCE COMICS

SPIDERS
Worldwide Webs

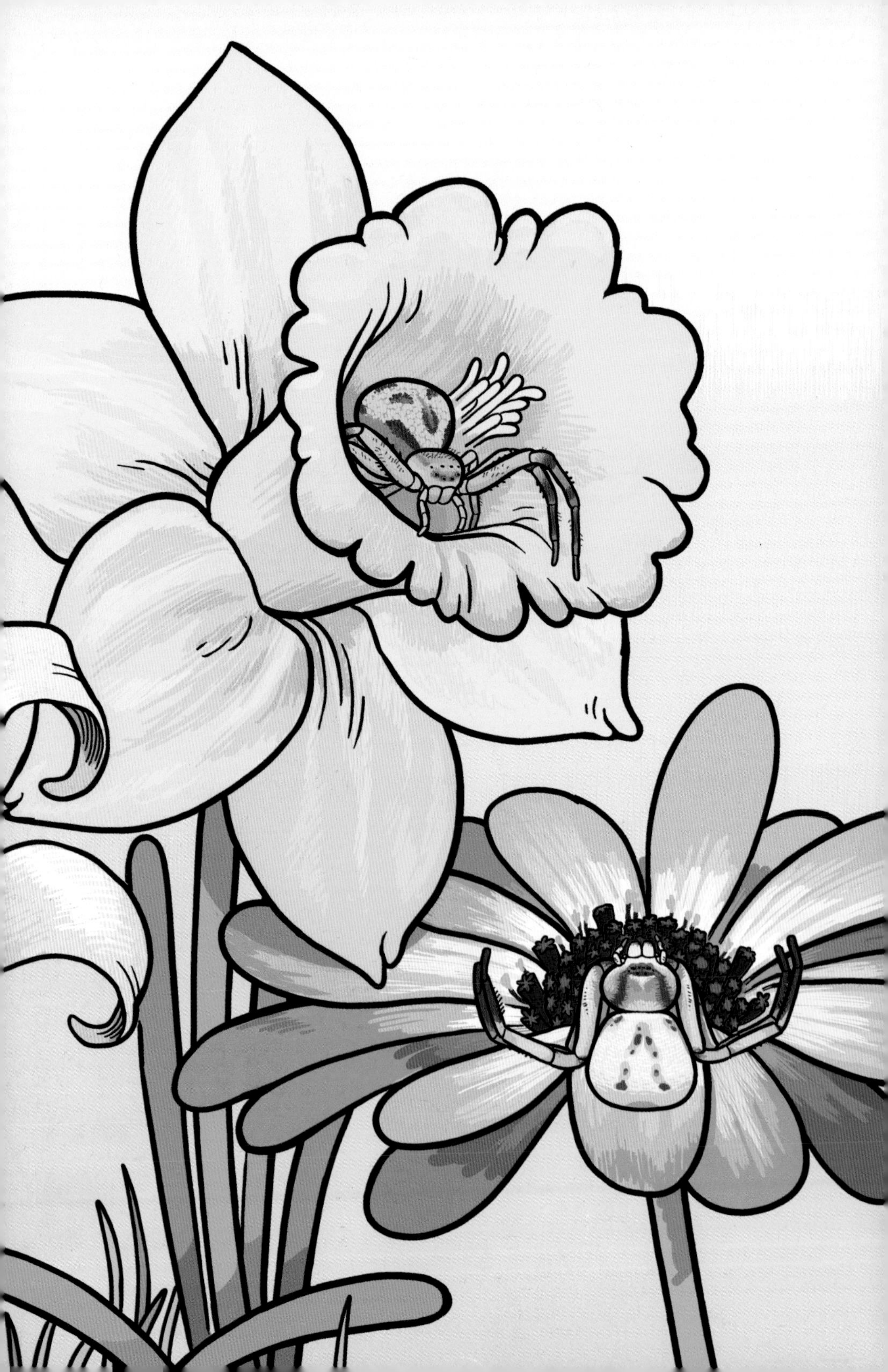

SPIDERS
Worldwide Webs

TAIT HOWARD

:01
First Second
New York

First Second

Published by First Second
First Second is an imprint of Roaring Brook Press,
a division of Holtzbrinck Publishing Holdings Limited Partnership
120 Broadway, New York, NY 10271

Don't miss your next favorite book from First Second! For the latest updates
go to firstsecondnewsletter.com and sign up for our enewsletter.

Library of Congress Control Number: 2020910689

Paperback ISBN: 978-1-250-22283-1
Hardcover ISBN: 978-1-250-22284-8

Our books may be purchased in bulk for promotional, educational, or business use. Please
contact your local bookseller or the Macmillan Corporate and Premium Sales Department
at (800) 221-7945 ext. 5442 or by email at MacmillanSpecialMarkets@macmillan.com.

First edition, 2021
Edited by Dave Roman
Cover and interior book design by Molly Johanson
Spiders consultant: Dr. Fiona Cross

Printed in China by Toppan Leefung Printing Ltd., Dongguan City, Guangdong Province

Penciled, inked, and colored digitally in Photoshop CC using a Wacom MobileStudio Pro
and a standard round Wacom pen nib.

Paperback: 10 9 8 7 6 5 4 3 2 1
Hardcover: 10 9 8 7 6 5 4 3 2 1

Spiders—they strike fear in some, fascination in others. Yell out "Spider!" in a room full of your friends and see what happens—half your friends will shriek in terror, and the other half will ask, "Where? Let me see it!"

When I was in college, I worked with researchers studying some gnarly trees growing on cliff faces, and while swinging from a climbing rope, I noticed that spiders were living a great life in the nooks and crannies of the cliffs. They were spinning big webs across sections of the rock and doing an amazing job of catching flies and other insects whizzing by. From that day on, I started to notice spiders everywhere, and I guarantee this will happen to you too! Our eight-legged friends are among the most common critters on the planet—they can be found living in forests or in washed-up seaweed on beaches, skating across small ponds while looking to catch fish, or even surviving on the slopes of Mount Everest. I've done my own research on Arctic spiders, and during the summer months in the far north, you can find wolf spiders every time you take a step—that's thousands of spiders per acre! In fact, arachnologists (the name for people who study spiders and their relatives) recently found that almost every home contains spiders, perhaps living in a dark closet or scurrying around the bathroom floor. I've done my own estimates too, and in most habitats, you are always within three feet of a spider. But don't be scared! Very few spider species can actually hurt us—their venom is most suited for catching smaller prey, often insects. In fact, in some croplands, spiders eat so many pest insects they actually help make sure farmers' fields are growing better food for us. They also eat insects that might hurt us or annoy us, such as mosquitoes. Some spiders eat other spiders too—taking over their web and devouring the resident.

Spiders are master weavers, but unlike Spider-Man (who uses high-tech webshooters attached to his wrists), spiders jet out their silken strands from specialized organs on their back ends. They use their threads to catch their prey, but they also use it to wrap up their eggs, or build protective shelters up in the corners of your bedroom or (as I learned!) on cliff faces. Some spiders can whip around a strand of silk (with a drop of glue at the end) and snag a moth—these are tiny little cowboys with silken lassos. Spiders also use silk to get airborne—they release threads upward, and their small bodies can be carried skyward, to be found thousands of feet in the air, before landing on isolated islands, perhaps in the middle of the Pacific Ocean. Even though we have known about this behavior for centuries, scientists are only now starting to understand how spiders do this "ballooning" (spoiler: it has to do with electrical currents!). Some spiders don't use silk to hunt; this includes spitting spiders, who hock a glue-like loogie at their prey.

And yet despite all the cool facts, some people are still scared of spiders, and a serious case of spider-fear is called "arachnophobia." Experts think such fear may be because we heard scary stories about spiders when we were children and it stuck with us. In some parts of the world, there are a few kinds of spiders that might bite humans and cause a reaction that may need a trip to the doctor, and arachnophobia may be an evolutionary adaptation rooted in these kinds of rare reactions. There's good news, though, because education is the gateway to turn fear into fascination, and the pages you are about to read will fill you with awe and wonder. You will learn about different kinds of spiders found around the world, and you will join Peter and Charlotte on a voyage of discovery, whether learning how silk is made, how spiders "balloon," or what amazing ways that spiders care for their young.

So, let's spin you a tale, and if you are a little worried, take a deep breath, open your mind, and be ready to learn about the planet's most incredible creatures. And the next time you are visiting a park, climbing around on some rocks, or waiting for a bus, keep your spidey sense on alert: an eight-legged friend is always close by, perhaps getting ready to sail the skies or catch a pest or two.

Professor Christopher Buddle
Arachnologist
McGill University, Montreal (Canada)

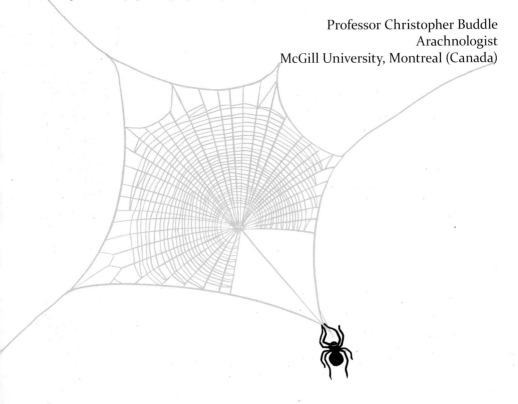

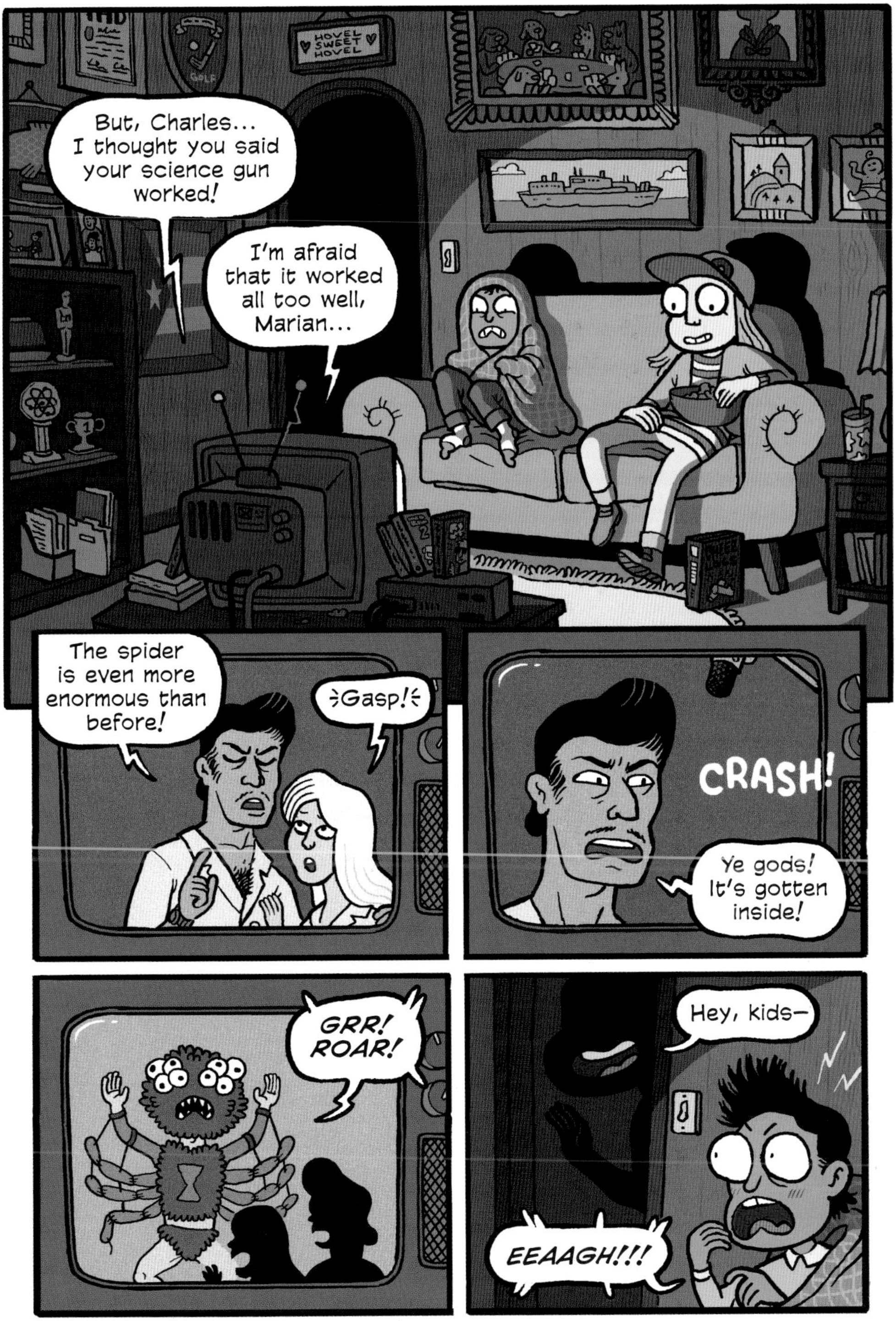
HOVEL SWEET HOVEL
But, Charles... I thought you said your science gun worked!
I'm afraid that it worked all too well, Marian...
The spider is even more enormous than before!
Gasp!
CRASH!
Ye gods! It's gotten inside!
GRR! ROAR!
Hey, kids—
EEAAGH!!!

Sorry! I need a favor. The university called, and I've gotta go in. Can you go down to the basement and find two boxes of laboratory glassware for me? I need them for tomorrow's classes.

click

‡huff‡ ‡huff‡

Sure thing, Mom! We'll get them now before we forget, right, Peter?

Huh? What are we doing?

Taking a trip... DOWNSTAIRS...

Don't torture your brother, Charlotte! I'll be back late, so eat whatever's in the fridge!

Wait! No! No no no!

Charlotte, stop! Please don't make me go down there!

Come on! It'll be good for you—a few little spiders aren't gonna hurt you!

What?! Of course they will! Don't you remember what happened at summer camp last year?!

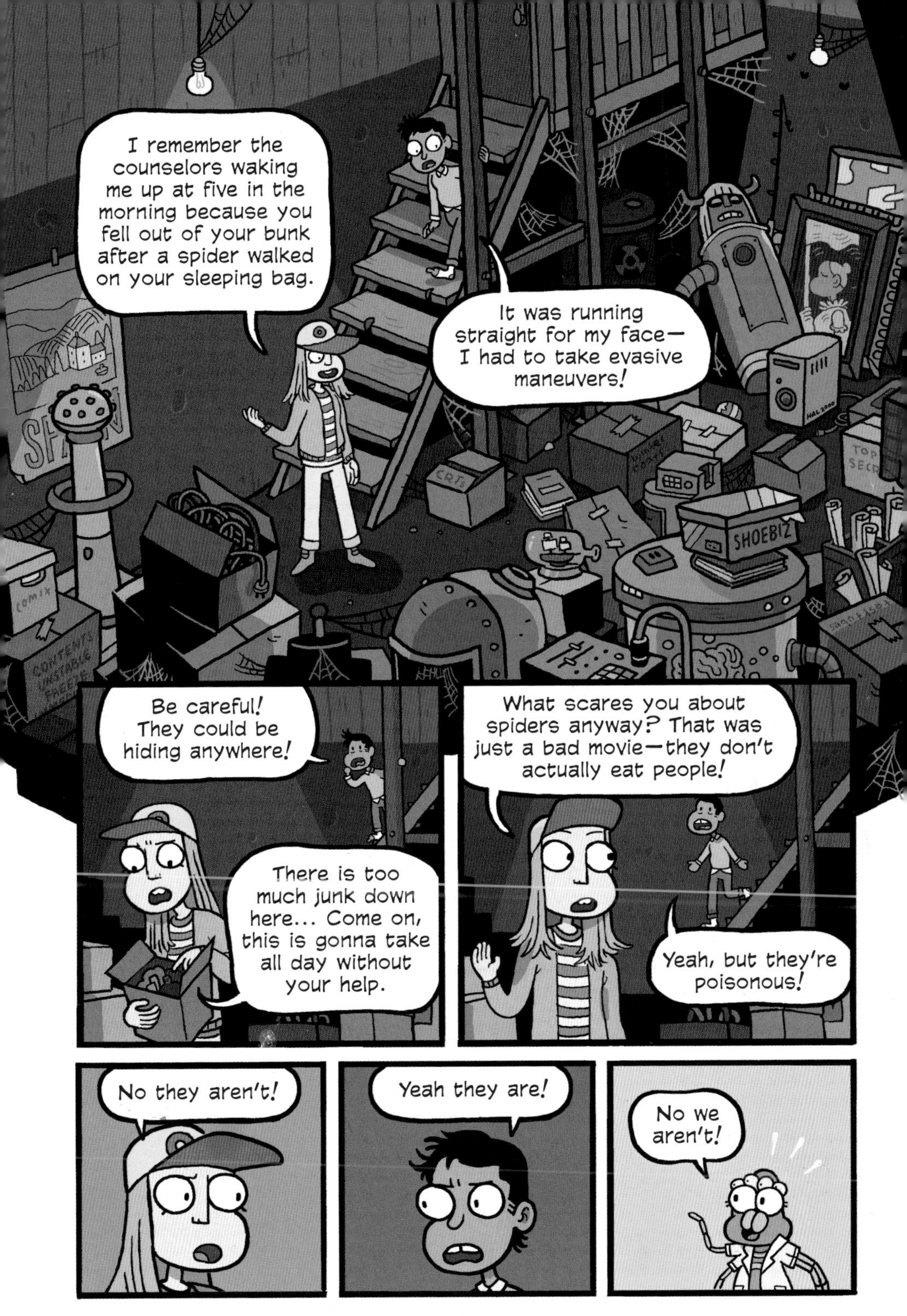
I remember the counselors waking me up at five in the morning because you fell out of your bunk after a spider walked on your sleeping bag.
It was running straight for my face—I had to take evasive maneuvers!
CRTs
SHOEBIZ
HAL 2000
TOP SECR
COMIX
CONTENTS UNSTABLE
Be careful! They could be hiding anywhere!
There is too much junk down here... Come on, this is gonna take all day without your help.
What scares you about spiders anyway? That was just a bad movie—they don't actually eat people!
Yeah, but they're poisonous!
No they aren't!
Yeah they are!
No we aren't!

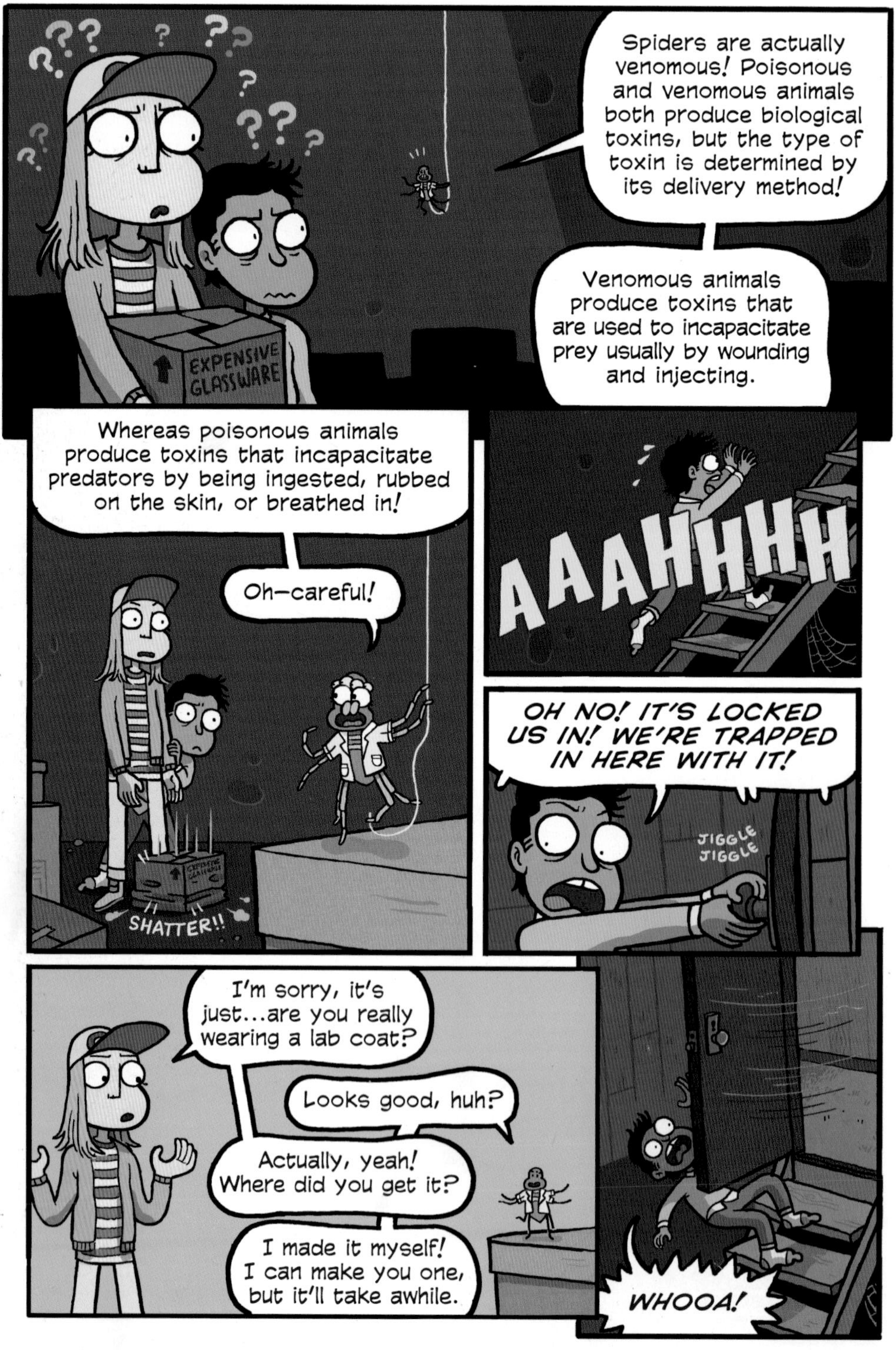

Spiders are actually venomous! Poisonous and venomous animals both produce biological toxins, but the type of toxin is determined by its delivery method!
Venomous animals produce toxins that are used to incapacitate prey usually by wounding and injecting.
EXPENSIVE GLASSWARE
Whereas poisonous animals produce toxins that incapacitate predators by being ingested, rubbed on the skin, or breathed in!
Oh—careful!
SHATTER!!
AAAHHHH
OH NO! IT'S LOCKED US IN! WE'RE TRAPPED IN HERE WITH IT!
JIGGLE JIGGLE
I'm sorry, it's just...are you really wearing a lab coat?
Looks good, huh?
Actually, yeah! Where did you get it?
I made it myself! I can make you one, but it'll take awhile.
WHOOA!

Urg!

VERY EXPENSIVE GLASSWARE

CRUNCH

Oh my! Are you okay?

Oooagh! I think so...

So what are you doing down here in our basement?

Well, we've gotta live somewhere too! Just ask any of the hundreds of spiders living in almost every home on earth!

...Hundreds?

Depending on the time of year and the size of the house, absolutely! Some spiders, like me, the American house spider, live here year round because we like to build webs in attics, basements, and other quiet dry places, but—

Oh gosh, I'm so sorry! I'm just going on and on without even introducing myself! I'm *Parasteatoda tepidariorum*, but you can call me Tepi!

That's a weird name.

PETER! Don't be rude just because she's not a human! That's her binomial name!

That's right! Binomial nomenclature is the formal system of naming species, usually using Greek or Latin words!

Since the beginning of language, people have categorized other living things. It has always been essential to know what kinds of plants you can eat and what kinds of animals might harm you!

But the modern scientific system of naming and defining groups of organisms by their physical traits is called taxonomy! Here's where I fit in the taxonomic ranking system!

KINGDOM: ANIMALIA

All animals on earth belong in this group! Most animals breathe oxygen, consume organic material, and can move around freely.

PHYLUM: EUARTHROPODA

Insects, arachnids, and crustaceans are all arthropods. They are invertebrates with an exoskeleton (which helps support their bodies and protect their organs), a segmented body, and pairs of jointed appendages.

CLASS: ARACHNIDA

All arachnids have eight legs, making them easy to distinguish from insects, which only have six. Besides spiders this class includes:

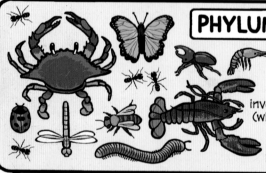

Scorpions

Ticks

Mites

Opiliones
(or "harvestmen")

Solifuges
(or "sunspiders," which aren't actually spiders!)

ORDER: ARANEAE

Spiders are the largest order of arachnids and one of the most diverse orders of all living organisms. They live on every continent except Antarctica and have colonized every environment except for the air and the open ocean!

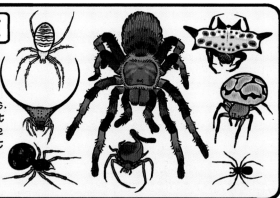

FAMILY: THERIDIIDAE

Also known as cobweb or tangle-web spiders. Even though other parasteatoda species are mostly found in Asia, the resilient theridiidae family can also be found across Europe and the Americas! In fact, it's the family you're most likely to encounter if you live in a home in North America!

GENUS: PARASTEATODA

A group made up of about 44 different species of mostly old-world spiders, meaning ones that originate in Europe, Asia, and Africa, as opposed to new-world spiders, which originate mostly in North or South America.

NEW WORLD   OLD WORLD

SPECIES: PARASTEATODA TEPIDARIORUM

FEMALE

MALE

That's me! This is the species anyone living in a home in North America is most likely to encounter. If you see cobwebs built in your attic or basement, they were probably built by one of these!

So where does the word "spider" come from? And how do you know how to build webs? And what is spider silk made of? How many kinds of spiders are there? And how do you know so much about spiders?

I've devoted myself to the lifelong pursuit of knowledge! I just happen to be the first ever *SPIDER BIOLOGIST!*

No you're not... We call them arachnologists, and they've been around for a while...

Fine! The first biologist spider! Now, the word "spider" comes from the Proto-Germanic word "spinþrô," which means "spinner." And—Oh! Wait! I've got an idea!

Do you two want to know more about spiders?

Yeah!

NOPE.

HALLOWEEN STUFF

Maybe if I teach you about spiders, you could both help me with something as well?

Wait, it's a trick! She just wants to drink our delicious blood!

You, shush! What can we do to help?

Oh, it's my baby—they're missing! I was in my web until early this morning catching breakfast, but when I went back to my lab, they were gone! All my other kids left the web a long time ago, but my sweet Maxie stayed to help with my research.

SOB SOB

If I don't find them, who will continue my work after I'm gone? Please, if you'll help me look for them, I'll tell you everything I know while we search!

No, I don't think we really have time to—

Of course we'll help! Let's go up to the attic first. I know I've seen a bunch of your friends around there.

Oh, no, just hold on—I've got a better idea.

Hmm... Where did I put... Ah!

TA-DA!

JUMBO SHRINK RAY

But if that's a shrink ray, then how did you shrink *IT?*

With that one over there! Now hold still!

Wait, but how did you shrink *THAT* one?

SCIENCE JUNK

ZAP

Ow, my skeleton!

WHODAA

Are you okay?

What...just happened...

Phew, I didn't know for sure this thing would work on living tissue!

Let's head to my lab, then I can introduce you to some of the spiders who live there!

Cool, let's meet a bunch of spiders. That'll help me be less afraid of spiders.

It actually can! The fear of spiders, arachnophobia, is one of the most commonly reported phobias on earth. There are a lot of reasons someone can be afraid of spiders, but learning about them, being exposed to pictures of them or to live ones in a safe, controlled environment can help some people overcome those fears!

CHOMP

TOYS

But the reason why it's so common isn't clear! It could be that your ancestors had this fear because spider ancestors posed a threat, so you developed an instinctual fear similar to a fear of heights.

But it's also likely that the wide-spread fear is due in part to social conditioning! It can be learned by watching other people be afraid of spiders or even by seeing their depiction in media as monsters!

A study at Ohio State University showed that when seeing a live spider, people who have a fear of spiders are much more likely to overestimate its size than people without that fear!

Even modern medicine can have a bias against spiders! There are almost forty medical conditions, including skin cancer and several deadly skin infections, that are often misdiagnosed by doctors as brown recluse spider bites.

Black widows also have a reputation as a "deadly" species, but there were no deaths that resulted from the 2,246 reported black widow bites in the USA in 2016.

The truth is that of the 48,353 known species of spider, only 200 or so can cause serious injury to humans.

Some species are more territorial, but spiders will only bite humans if threatened, and they won't always inject a full dose or even any venom at all when making a defensive bite.

KEEP OUT

Any spider's bite can be painful, but spider fangs are usually meant for biting into insects!

For small spiders, trying to bite skin would be like you trying to take a bite out of a wall! Some larger spiders, like tarantulas, have fangs that cause a more painful bite, but tarantula venom isn't actually dangerous to humans!

FRUITFLY JELLY

1 day

1 week

1.5 weeks

2-4 weeks

GNAW GNAW GNAW

Whoa, this is hard!

Why are some so much more venomous than others?

Different kinds of spiders each have unique venom that affects other animals differently. We usually evolve venoms that suit our prey. Most spiders eat just insects and other spiders...but larger spiders can take down much larger creatures!

THE **Spider Diet**

Insects

Spiders

Rodents

Lizards

Frogs

Small Birds

Fish

S.S. PIDER

Wait, spiders can *fish?!*

Some can, yeah! And I'll introduce you to them too, but first let's meet a few of the other spiders who share your house!

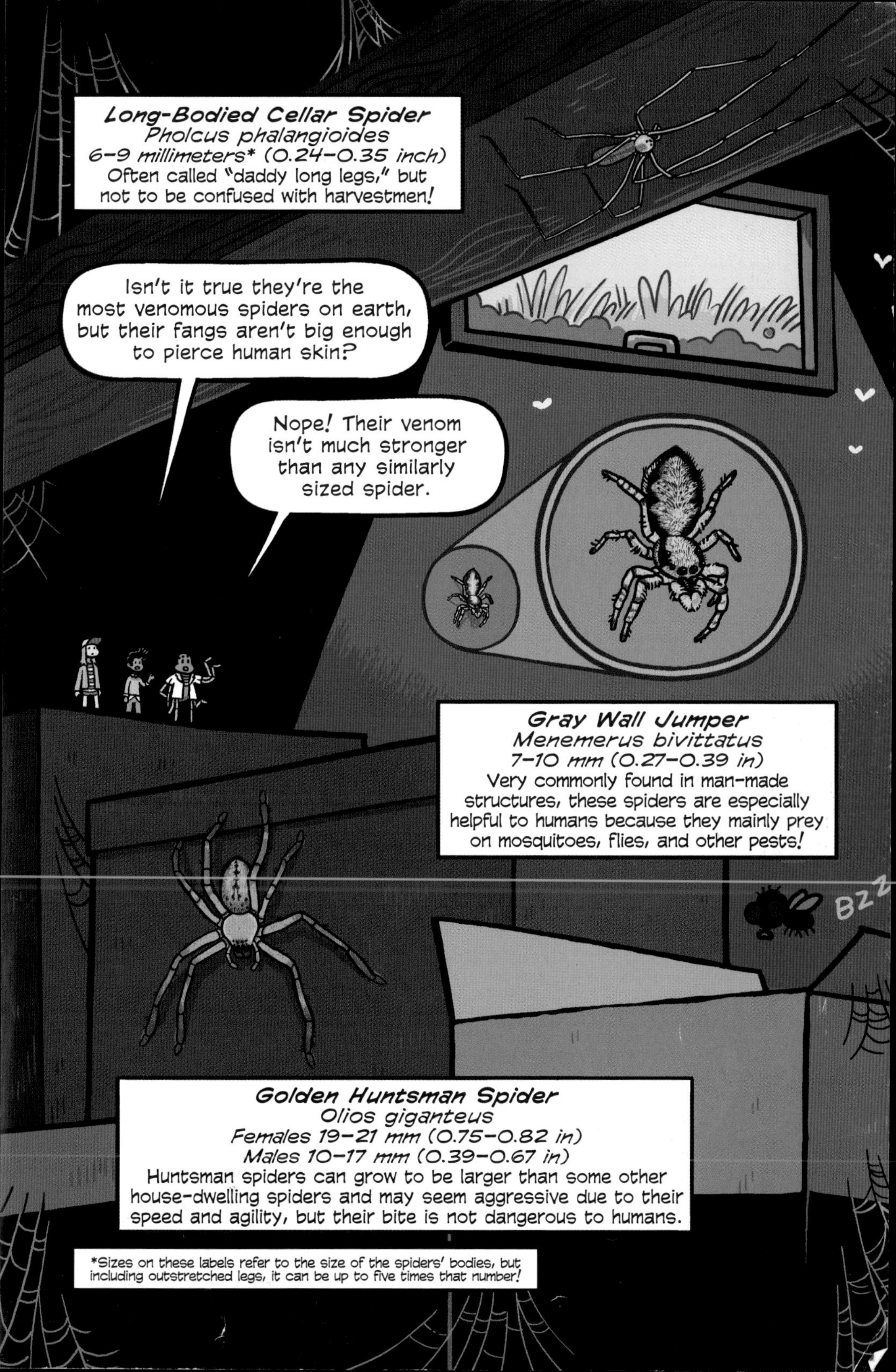

Long-Bodied Cellar Spider
Pholcus phalangioides
6–9 millimeters* (0.24–0.35 inch)
Often called "daddy long legs," but not to be confused with harvestmen!
Isn't it true they're the most venomous spiders on earth, but their fangs aren't big enough to pierce human skin?
Nope! Their venom isn't much stronger than any similarly sized spider.
Gray Wall Jumper
Menemerus bivittatus
7–10 mm (0.27–0.39 in)
Very commonly found in man-made structures, these spiders are especially helpful to humans because they mainly prey on mosquitoes, flies, and other pests!
BZZ
Golden Huntsman Spider
Olios giganteus
Females 19–21 mm (0.75–0.82 in)
Males 10–17 mm (0.39–0.67 in)
Huntsman spiders can grow to be larger than some other house-dwelling spiders and may seem aggressive due to their speed and agility, but their bite is not dangerous to humans.
*Sizes on these labels refer to the size of the spiders' bodies, but including outstretched legs, it can be up to five times that number!

FEMALE

Howdy!

Spots on their abdomen above their spinnerets.

MALE

Southern Black Widow
Latrodectus mactans
Females 8–13 mm (0.31–0.51 in)
Males 3–6 mm (0.12–0.24 in)
These spiders get a bad reputation because of the highly toxic nature of their venom, but they rarely leave their webs and are not inclined to bite humans. Females have a red hourglass mark on the underside of their abdomen, which is sometimes split in the middle.

Oh, hey, Tepi! Are these friends of yours or lunch for later?

They're helping me look for Max!

Aww, Maxie ran off again, huh? I think I saw them out by the garden, but—

14

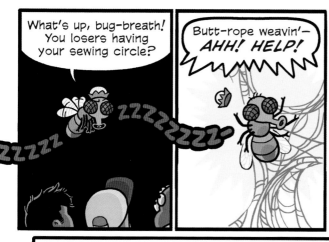

Black widows, like many web-weaving spiders, begin capturing prey by wrapping it in thick bands of silk, using their front legs to spin the prey rapidly and back legs to pull the silk. This process helps immobilize prey quickly!

WHOᴼᴼᴼᴀ!!

Theridiidae spiders like the black widow are sometimes called comb-footed spiders because of a special row of bristles on their hind "feet" that help them spread silk quickly over prey.

For more dangerous prey, like wasps or crickets, the wrapping process only takes a second or two as the spider tries to avoid being bitten or stung! This speed also comes in handy when multiple victims are in the web at the same time!

Why I oughta!

To bite prey, the spider uses their fangs, which are part of its front appendages, called chelicerae! ("kuh-lis-er-ee")

The fangs are articulated and normally rest in a small groove in the chelicerae like the blade of a folding knife.

Serrated edge for cutting silk threads or mashing up food.

The opening to inject venom is just above the tip of the fang. This prevents clogging and makes the tip less likely to be damaged when biting into prey.

Most modern spiders fall into two distinct categories, partially based on the way their chelicerae move!

Araneomorphae

- 93–94% of modern spiders
- Usually smaller in size
- Shorter lifespan, 1–3 years
- Chelicerae move sideways

Mygalomorphae

- Usually larger: tarantulas, etc.
- Can live for up to 25 years
- Chelicerae move outward

The oldest spider ever recorded, an Australian trapdoor spider, lived to be 43!

Chelicerae are also useful for things like digging or carrying eggs. The assassin spiders use their elongated chelicerae to spear prey!

Venom is pushed out by powerful muscles surrounding the venom glands, which lie either inside the chelicerae or extend deeper into the head and thorax, also known as the cephalothorax.

Uurg...

Once the venom is injected, it takes effect quickly.

Brain

Venom Gland

Mouth Opening & Esophagus

Spider venoms can contain neurotoxins, which affect the central nervous system and either immobilize or kill prey depending on the kind of venom and strength of the dose delivered.

They can also contain cytotoxins, which help break down the internal tissues and organs of their prey to make it more digestible.

Using a special sucking stomach, the spider sucks the liquefied innards out of the insect, filtering out anything too big to swallow with bristles surrounding their mouth.

YUM!

Adult spiders also have intestinal systems that allow them to go a long time without eating. A black widow can go months without food as long as it has water!

SLURP SLURP

BUG GULP

But we also get a lot of water from our food as well!

So they wrap it to make sure it can't attack while they inject venom?

Yup! Some spiders bite before wrapping, but it depends on the type of spider, the prey, and the situation! For smaller prey, only a few small threads of silk may be needed to immobilize it.

Pedipalps, the smaller appendages on the front of the spider, are extremely versatile. They help the spider capture and manipulate prey, dig, carry eggs, and much more. They also play an important role in mating and help humans distinguish males from females.

NUM NUM NUM

Do spiders that don't use webs to hunt still make silk?

All spiders produce silk. Some just use it differently! Let's search the garden so I can introduce you to some spiders there who know all there is to know about silk!

What's so special about spider silk? Aren't there worms that make silk as well?

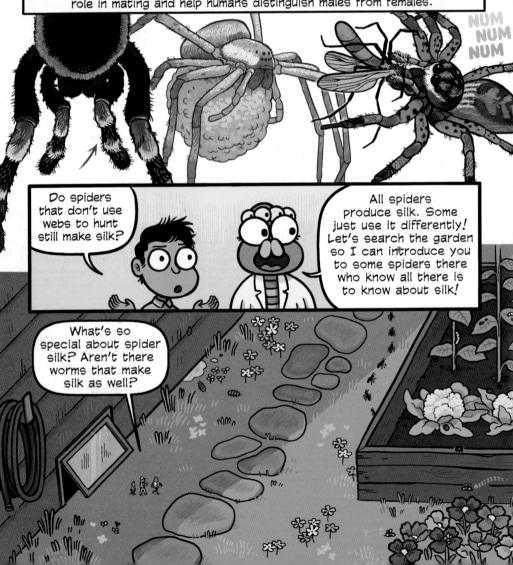

Yup! Butterflies, moths, bees, beetles, fleas, mites, crickets, and some ants make silk too, but usually only one kind during the larval stages of their life.

A spider, on the other hand, has appendages called spinnerets on their abdomen that allow them to make up to eight different types of silk.

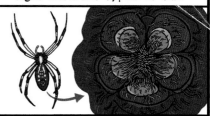

Silkworms are the number one producer of silk used in fabric throughout the world! Their silk is very similar to spiders', but they produce large quantities all at once to build cocoons, which can be boiled to release the silk fibers.

But spider silk has a few advantages!

Its stickiness and elasticity could lead to stitches that help a wound close better.

Because of its strength, it could be used to make artificial muscles!

It's already being used to improve the effectiveness of car airbags and bulletproof vests!

Modern means of extracting spider silk are far from efficient though! It would take about a million spiders and a team of seventy people *two years* to produce a postcard-sized piece of fabric made from spider silk!

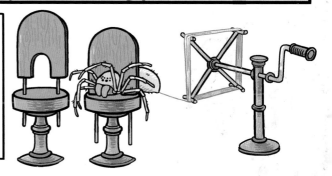

Silkworms mostly eat mulberry leaves, so keeping a lot of them together isn't a problem, but spiders, well...

They eat other spiders! This makes keeping large quantities of them for farming silk difficult!

Hey, cut it out!

CHOMP CHOMP

When researchers at the University of Wyoming tried to use a big group of spiders to produce silk for research, the territorial spiders either ate or killed one another.

But they realized that the solution was simple:

GOATS!

CHEW CHEW CHEW

Uh...goats?

Yup! GOATS!

Genetic modification is a process used to produce desired traits in plants or animals, and humans have been doing it for over ten thousand years. The most basic form of this is selective breeding—picking out "superior" individuals, which are used to breed offspring that inherit "superior" traits!

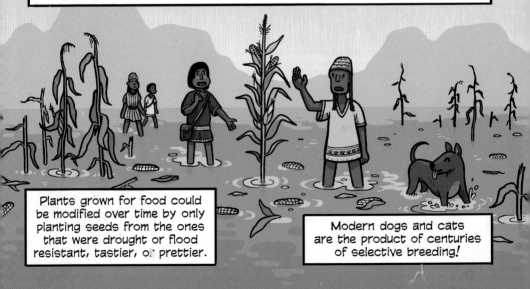

Plants grown for food could be modified over time by only planting seeds from the ones that were drought or flood resistant, tastier, or prettier.

Modern dogs and cats are the product of centuries of selective breeding!

Today there are more precise ways to genetically modify organisms! Using a process called transfection (transfer + infection), foreign DNA can be inserted into the nucleus of a cell! A few methods of doing this are:

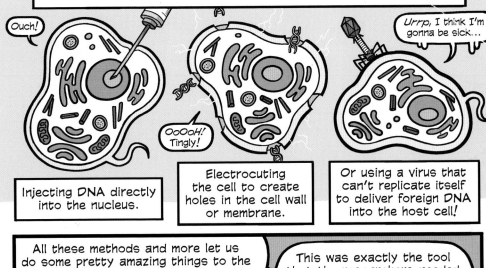

Ouch!

OoOoH! Tingly!

Urrp, I think I'm gonna be sick...

Injecting DNA directly into the nucleus.

Electrocuting the cell to create holes in the cell wall or membrane.

Or using a virus that can't replicate itself to deliver foreign DNA into the host cell!

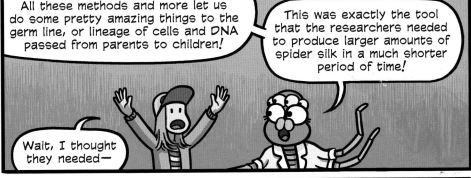

All these methods and more let us do some pretty amazing things to the germ line, or lineage of cells and DNA passed from parents to children!

This was exactly the tool that the researchers needed to produce larger amounts of spider silk in a much shorter period of time!

Wait, I thought they needed—

Goats? They needed both! By introducing some of the silk-producing genes of a golden orb-weaver into a goat embryo, they created goats that look and act like normal goats, but whose milk contains spider-silk proteins!

SPIDER MILK

SPIDER MILK

By separating the proteins from the milk, we can make spider silk faster and more efficiently than ever before!

HISSS

Whoa...

Has your silk always been "his" important?

To us it sure has! It's been essential to our evolutionary success over all these years!

If we check out the fossil record, it can tell us when and why spiders evolved silk! Intact spider fossils are rare, but over time we've put together pieces of our long evolutionary history!

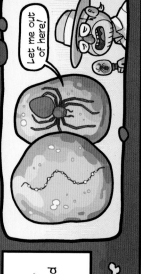

Let me out of here!

130 MILLION YEARS AGO: Oldest sticky thread in amber. This is the earliest example of preserved silk that resembles a modern spider's. When the sap or resin from a tree fossilizes, we call it amber! Spider fossils are often hard to study because they're usually only found in fragments, but amber can preserve the whole specimen as well as DNA because it traps and dehydrates the organism!

225 MILLION YEARS AGO: Oldest araneomorph fossil. These were some of the first spiders that were able to produce major ampullate silk (dragline silk). Ampullate silk is what allowed these spiders to build the first webs, as it was strong enough to support the weight of the spider or its prey!

240 MILLION YEARS AGO: Oldest mygalomorph fossil. The first spiders in this order were similar to modern tarantulas! The greater abundance of plants that could protect them from the harsh sun also led to more abundant prey outside of their burrows, so they began to construct webs leading outside.

290 MILLION YEARS AGO: Oldest true spider fossil. Mygalomorphs and araneomorphs make up 99.9% of all living spiders today, but the third group, Mesothelae, are the closest relatives to these first true spiders. These spiders had two or four spinnerets in the middle of their abdomen. They used their silk to line their burrows, which reinforced the walls and helped regulate the temperature within.

305 MILLION YEARS AGO: Oldest *Idmonarachne* fossil. *Idmonarachne brasieri* is named after the father of Arachne in the Greek myth that gave arachnids their name! These spiderlike arthropods still had a segmented abdomen and still had no spinnerets, but it's easy to see how that segmented backside eventually became the smooth one spiders have today!

380 MILLION YEARS AGO: Oldest *Attercopus* fossil. *Attercopus fimbriunguis* is one of the spider's oldest relatives. Systematists, who classify organisms into taxonomical ranks by studying their evolution and relationships with other animals, originally classified it as the oldest spider ever, but later they realized its silk glands weren't true spinnerets.

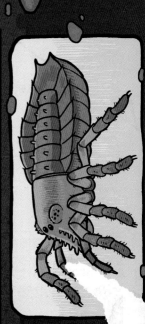

419 MILLION YEARS AGO: Oldest trigonotarbid fossil. Although they are more closely related to modern harvestmen than spiders, these early arachnids share a few distinct characteristics with modern spiders, like their eight legs and one pair of pedipalps, two-part body, and pocketknife-style fang-tipped chelicerae, although they had no venom glands.

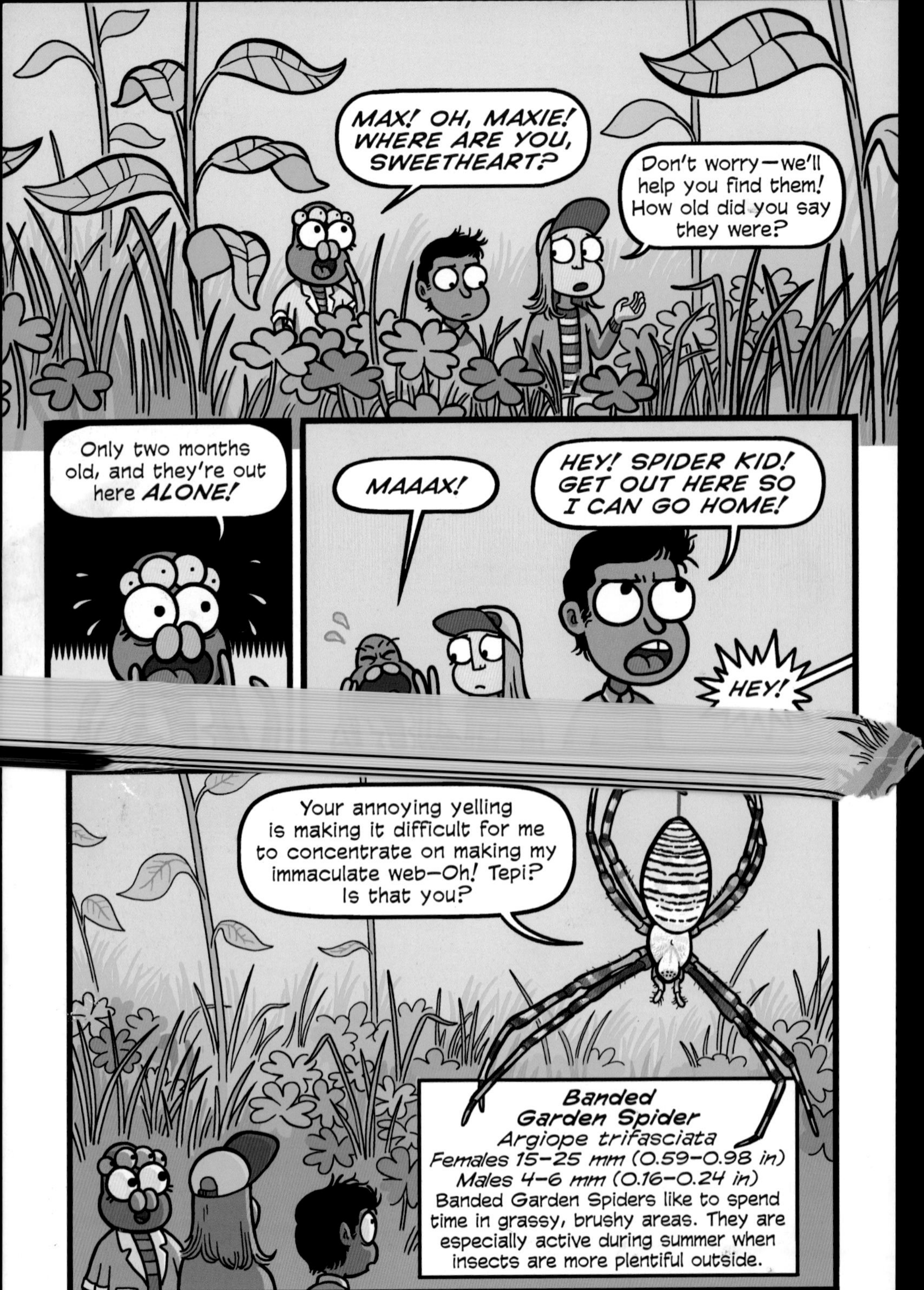
MAX! OH, MAXIE! WHERE ARE YOU, SWEETHEART?
Don't worry—we'll help you find them! How old did you say they were?
Only two months old, and they're out here ALONE!
MAAAX!
HEY! SPIDER KID! GET OUT HERE SO I CAN GO HOME!
HEY!
Your annoying yelling is making it difficult for me to concentrate on making my immaculate web—Oh! Tepi? Is that you?
Banded Garden Spider
Argiope trifasciata
Females 15–25 mm (0.59–0.98 in)
Males 4–6 mm (0.16–0.24 in)
Banded Garden Spiders like to spend time in grassy, brushy areas. They are especially active during summer when insects are more plentiful outside.

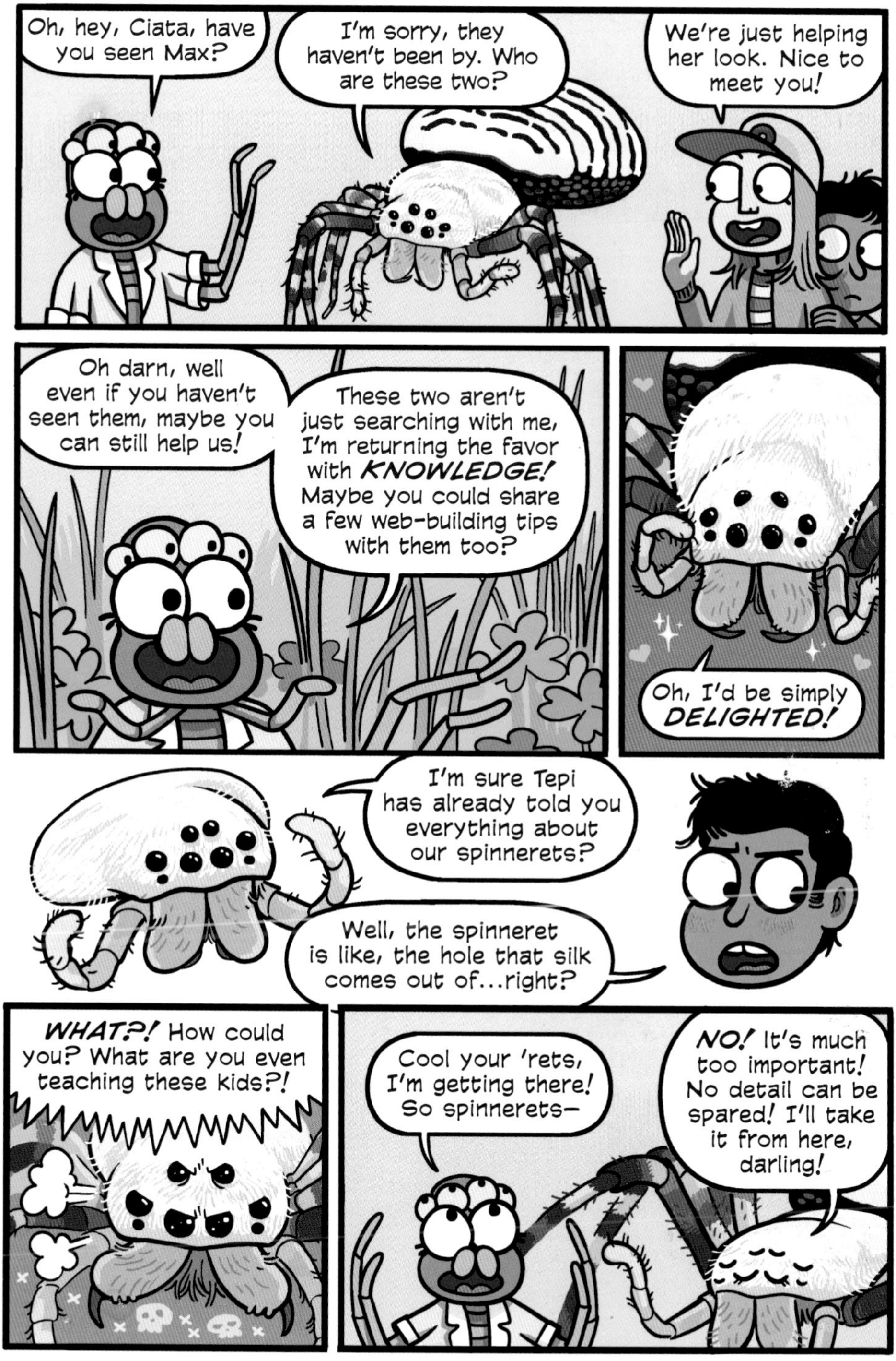
Oh, hey, Ciata, have you seen Max?
I'm sorry, they haven't been by. Who are these two?
We're just helping her look. Nice to meet you!
Oh darn, well even if you haven't seen them, maybe you can still help us!
These two aren't just searching with me, I'm returning the favor with KNOWLEDGE! Maybe you could share a few web-building tips with them too?
Oh, I'd be simply DELIGHTED!
I'm sure Tepi has already told you everything about our spinnerets?
Well, the spinneret is like, the hole that silk comes out of...right?
WHAT?! How could you? What are you even teaching these kids?!
Cool your 'rets, I'm getting there! So spinnerets—
NO! It's much too important! No detail can be spared! I'll take it from here, darling!

Major & Minor Ampullate Glands

Major ampullate silk is the base for most webs. It's also used for the dragline, which many spiders leave behind them everywhere, allowing them to stop a fall or retreat to their burrow. Minor ampullate threads are used while building webs as well.

Piriform Glands

These glands produce tiny discs of sticky silk, which are used to glue threads in a web together to improve stability. They harden on contact with the air and can stick to a wide variety of materials. These discs also serve as the anchors for dragline threads!

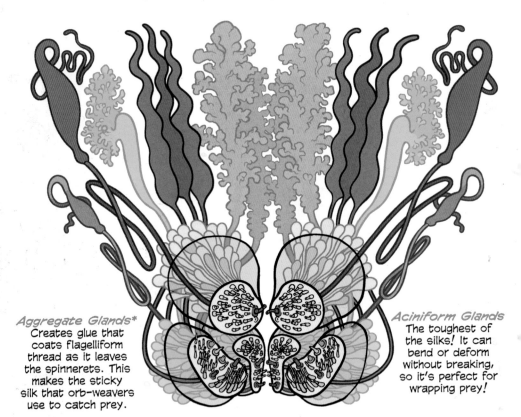

Aggregate Glands*

Creates glue that coats flagelliform thread as it leaves the spinnerets. This makes the sticky silk that orb-weavers use to catch prey.

Aciniform Glands

The toughest of the silks! It can bend or deform without breaking, so it's perfect for wrapping prey!

Flagelliform Glands*

The "capture spiral" of a web is made of this extremely elastic silk. When something flies into the web, it can stretch far, giving the glue longer to stick to the insect.

Tubuliform Glands

Most modern female spiders make their egg sacs from this silk! This is the stiffest kind of silk, meaning it can withstand more direct force to better protect the eggs inside.

**Only orb-weavers have these two glands!

But when we make any of those silks, they're formed inside us as a liquid! It's the special shape of the spider's silk glands as well as their spinnerets that allow it to be formed into a solid thread!

Silk is mostly proteins, which are made up of long chains of amino acids. Under a powerful microscope, they look like a bunch of small coiled-up springs!

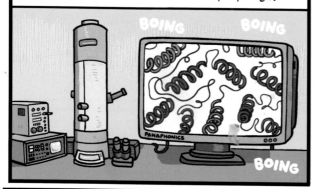

BOING BOING BOING

PANAPHONICS

Liquid proteins are stored in these glands until needed, when they are then forced rapidly through a thin duct with tight turns. This is where the magic happens!

The lining of this long structure removes some water from the liquid silk, and the thin squeeze orients the proteins into a long structured fiber. This realignment of proteins is what gives spider silk its strength and durability!

Move it!

Outta my way!

Each spigot has a valve that regulates the amount of silk leaving the gland or clamps down to stop us from falling.

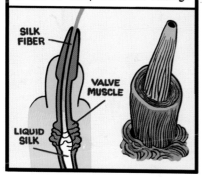

SILK FIBER

VALVE MUSCLE

LIQUID SILK

Clusters of spigots make up each spinneret. The spigots make tiny strands of silk that fuse with the ones nearby, and the weight of the spider or the pulling motion of their foot extrudes the liquid silk into a solid thread!

Most spiders have six spinnerets laid out in three rows of two located on the end of their abdomen.

Mesothelae

Tarantula

Golden Silk Orb-Weaver

But they can have between two and eight! Orb-weavers tend to have more, as they produce a wider variety of silks.

They're movable too! Each spinneret has muscles surrounding it, which allow it to move in tandem with others or independently, giving the spider a lot of control when laying down silk.

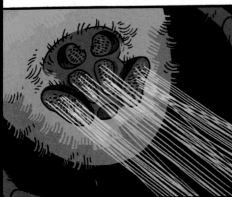

Liphistiidae, the only living family of Mesothelae spiders, has the same spinnerets as its ancient relatives. These are located on the bottom side of their abdomen and are mostly used to line their burrows with silk.

Who would've thought so much was going on in that giant butt!

How *RUDE!*

Hold up. You said there were eight kinds of silk, but you only told us about seven of them!

HAH! No detail spared? I can answer this one!

Cribellate spiders have a special organ located above their spinnerets called a cribellum, made up of one or more plates covered in thousands of microscopic holes.

THIS THING!

These holes produce fine strands of silk that are brushed out by a special comb on their back feet called a calamistrum. This process results in a woolly silk that forms a powerful adhesive that doesn't dry out!

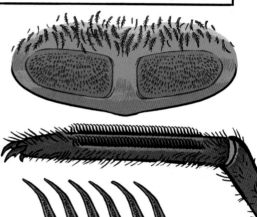

Fine, fine! I forgot one thing! I can't be expected to remember *every* detail of every old spider on earth, but I can teach you to build nature's most amazing achievement—

THE ORB WEB!

Why do they call it that?

Because the word "orb" used to also mean "ring" or "circle," and the name has stuck for hundreds of years!

HOW to MAKE

After reaching a high enough point, the spider releases lines of silk that "balloon" out.

FSHHH

1

It may take several tries to find the right spot, but once it attaches to a suitable anchor point, she can begin! The first strand is called the bridge thread, and it can also be made by walking from one point to another.

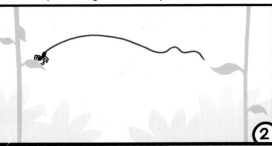

2

She connects the anchor points, then makes smaller framing lines around the inside of this shape. This structure is a frame for the spokes of the web.

5

She travels back and forth between the outside and center of the web to build the spokes, or "radii"—the support structure for sticky thread.

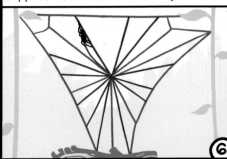

6

The capture thread is tightly spiraled to create the "net" that will catch fast-moving prey.

9

As she lays the capture thread, she eats the temporary silk, which her body can reuse later.

SSSLURP

10

The aggregate glue coats the capture thread evenly—

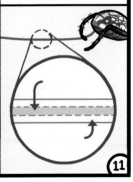

11

an ORB WEB!

She doubles back with a looser thread and then cuts the bridge thread in the middle by biting it.

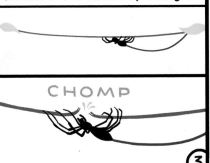

CHOMP

(3)

She drops from the middle and anchors the looser thread to the ground, creating a **Y** shape that forms the base of the web.

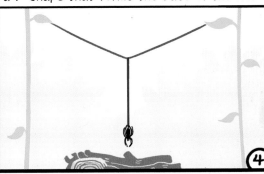

(4)

The first spiral on the web is made with minor ampullate silk. This is temporary thread that is used to stabilize the web.

(7)

She releases flagelliform silk coated with aggregate glue. This forms the capture thread that prey sticks to.

THWIP

(8)

—so at the end of each section, she flicks the thread, breaking up the glue into smaller droplets, which will bond to prey much faster.

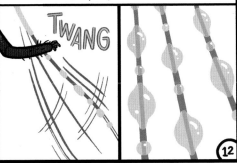

TWANG

(12)

And there you have it! Orb-weavers often hang head down in the center of the web, using vibrations to tell when prey has been caught.

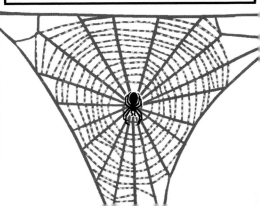

31

So then what stops you from getting stuck in your own web?

A few things prevent us from getting tangled up!

Spiders' legs are covered in tiny hair-like bristles called setae, which don't have enough surface area for the glue to form a bond with.

Spiders move carefully to prevent the web from bouncing back at them. They also walk on nonsticky threads whenever possible.

Setae

BUT IT'S PRONOUNCED **See-tee**

Some web-building spiders have three claws on each leg: two combed ones on either side of a small hook that grasps strands of silk by pressing against serrated bristles across from the claws. The thread gets caught on these notches, and the spider can release the silk by lifting the claw back up.

My word!

SPROING

BOUNCE!

A spider walking along a thread might brush against one or two glue droplets at a time, but a fly hitting a web can make contact with about fifty!

And finally, a special nonstick chemical produced by the spider coats their legs, which aids them in crossing the web unencumbered!

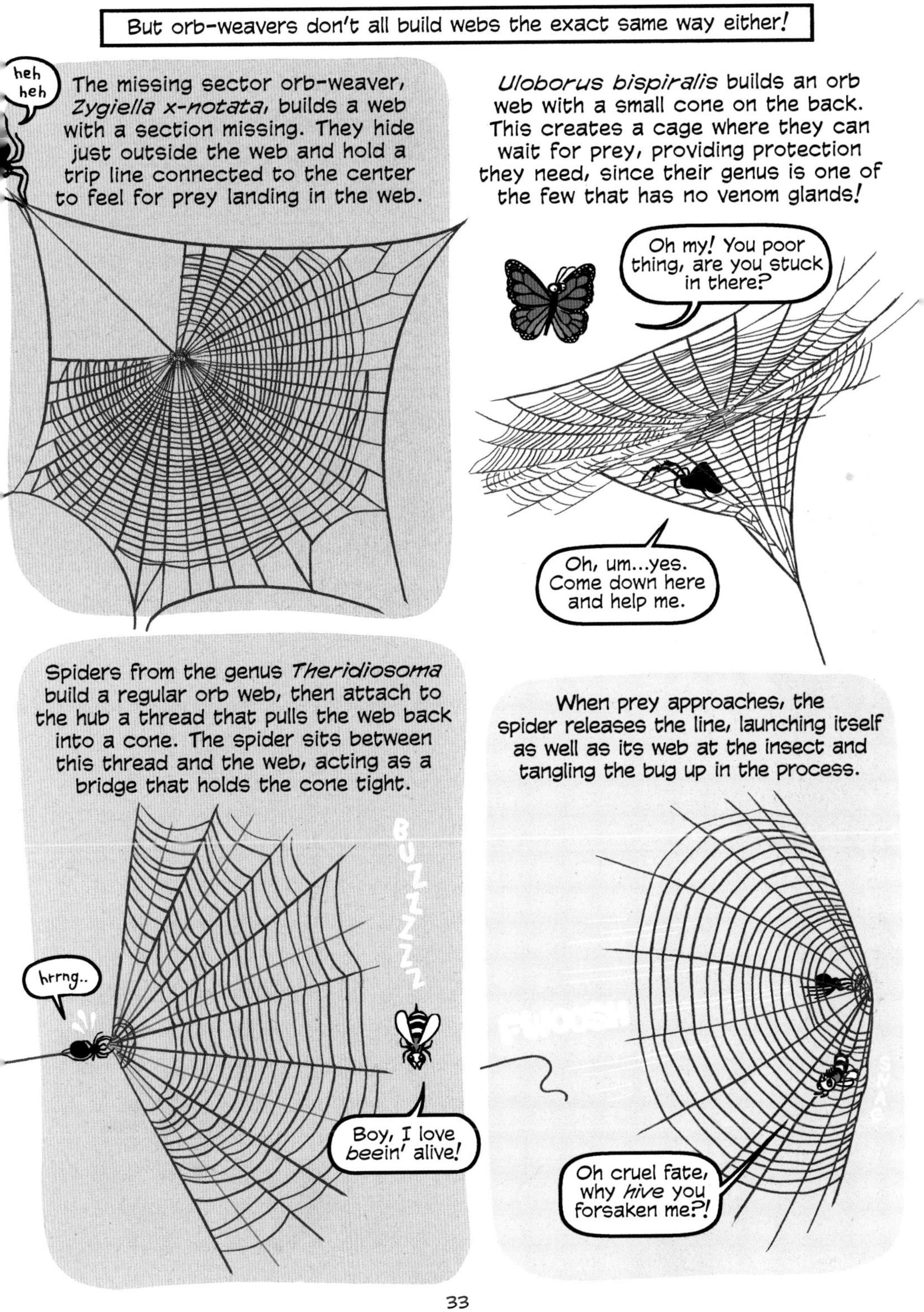
But orb-weavers don't all build webs the exact same way either!
heh heh
The missing sector orb-weaver, *Zygiella x-notata*, builds a web with a section missing. They hide just outside the web and hold a trip line connected to the center to feel for prey landing in the web.
Uloborus bispiralis builds an orb web with a small cone on the back. This creates a cage where they can wait for prey, providing protection they need, since their genus is one of the few that has no venom glands!
Oh my! You poor thing, are you stuck in there?
Oh, um...yes. Come down here and help me.
Spiders from the genus *Theridiosoma* build a regular orb web, then attach to the hub a thread that pulls the web back into a cone. The spider sits between this thread and the web, acting as a bridge that holds the cone tight.
hrrng..
BUZZZZZ
Boy, I love *beein'* alive!
When prey approaches, the spider releases the line, launching itself as well as its web at the insect and tangling the bug up in the process.
FWOOSH
Oh cruel fate, why *hive* you forsaken me?!

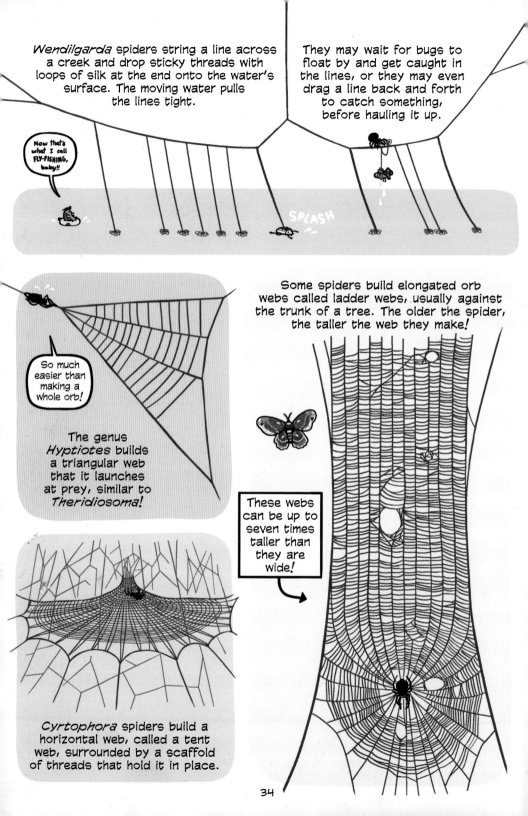

Wendilgarda spiders string a line across a creek and drop sticky threads with loops of silk at the end onto the water's surface. The moving water pulls the lines tight.

They may wait for bugs to float by and get caught in the lines, or they may even drag a line back and forth to catch something, before hauling it up.

Now that's what I call FLY-FISHING, baby!!

SPLASH

So much easier than making a whole orb!

The genus *Hyptiotes* builds a triangular web that it launches at prey, similar to *Theridiosoma!*

Some spiders build elongated orb webs called ladder webs, usually against the trunk of a tree. The older the spider, the taller the web they make!

These webs can be up to seven times taller than they are wide!

Cyrtophora spiders build a horizontal web, called a tent web, surrounded by a scaffold of threads that hold it in place.

34

And those are just a handful of examples! There are over three thousand species of orb-weavers, each with its own—

Yes yes, this is all very nice, but if you'll excuse me— it's lunchtime!

HELP!

Oh! Thank goodness! This invisible net seems to have gotten the best of me. If you could just help—

Oh, I'm quite warm. There's no need to wrap me up—*Oh!* What are you—*Oh no!*

THOSE ARE MY ORGANS! RETURN THEM AT ONCE!

What happens when your web is destroyed? Do you just leave it and start over?

Nope, we eat it!

Orb-weavers may build a few webs each day, so to conserve energy, they eat any unused silk, and their digestive system sends the proteins back to their silk glands to reuse later!

Yum! It's *me* flavored!

NUM NUM NUM NUM NUM NUM NUM NUM

Orb-weavers usually take their webs down each night and rebuild them the next morning!

Ah, right through here—

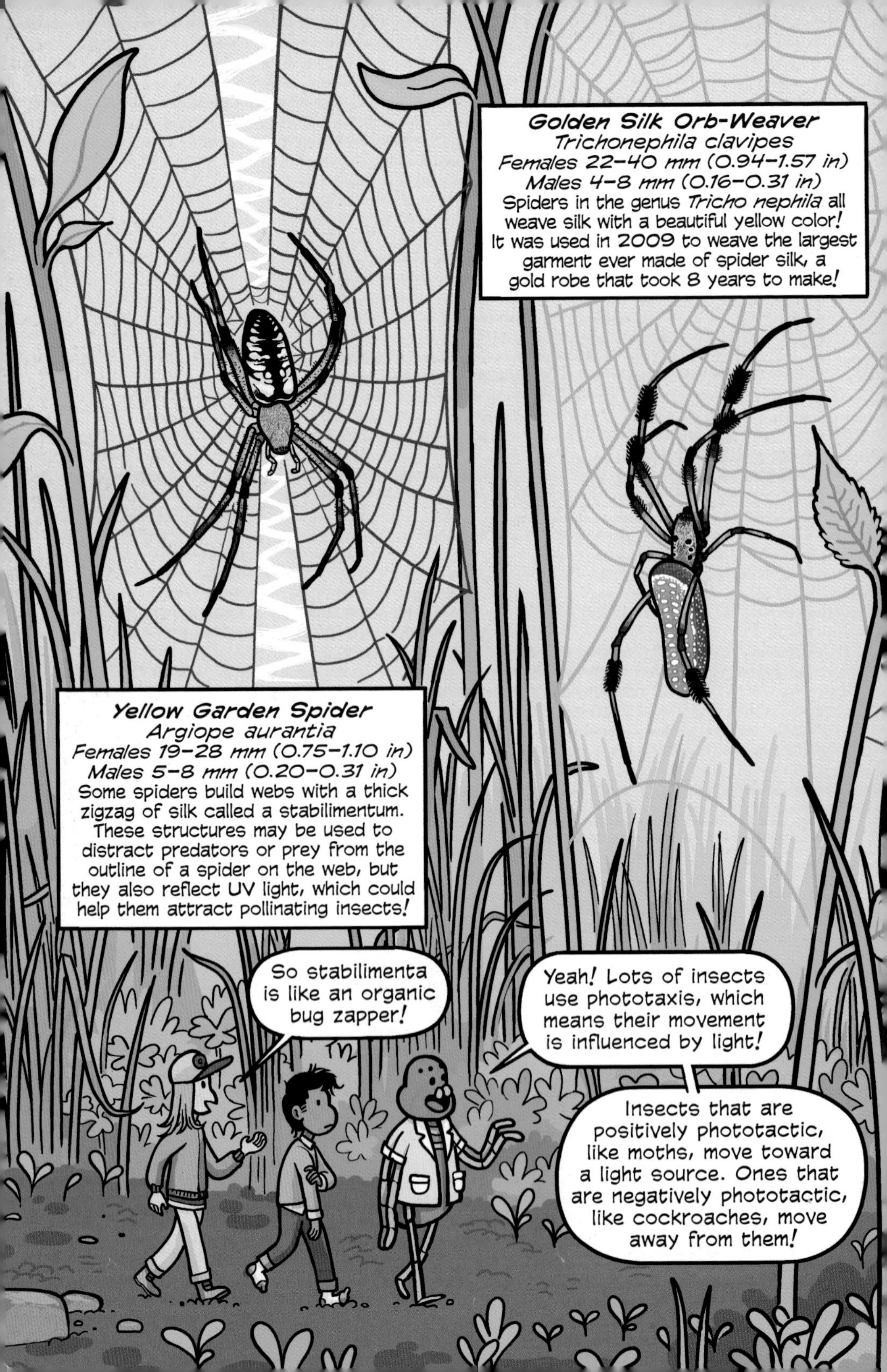
Golden Silk Orb-Weaver
Trichonephila clavipes
Females 22–40 mm (0.94–1.57 in)
Males 4–8 mm (0.16–0.31 in)
Spiders in the genus Tricho nephila all weave silk with a beautiful yellow color! It was used in 2009 to weave the largest garment ever made of spider silk, a gold robe that took 8 years to make!
Yellow Garden Spider
Argiope aurantia
Females 19–28 mm (0.75–1.10 in)
Males 5–8 mm (0.20–0.31 in)
Some spiders build webs with a thick zigzag of silk called a stabilimentum. These structures may be used to distract predators or prey from the outline of a spider on the web, but they also reflect UV light, which could help them attract pollinating insects!
So stabilimenta is like an organic bug zapper!
Yeah! Lots of insects use phototaxis, which means their movement is influenced by light!
Insects that are positively phototactic, like moths, move toward a light source. Ones that are negatively phototactic, like cockroaches, move away from them!

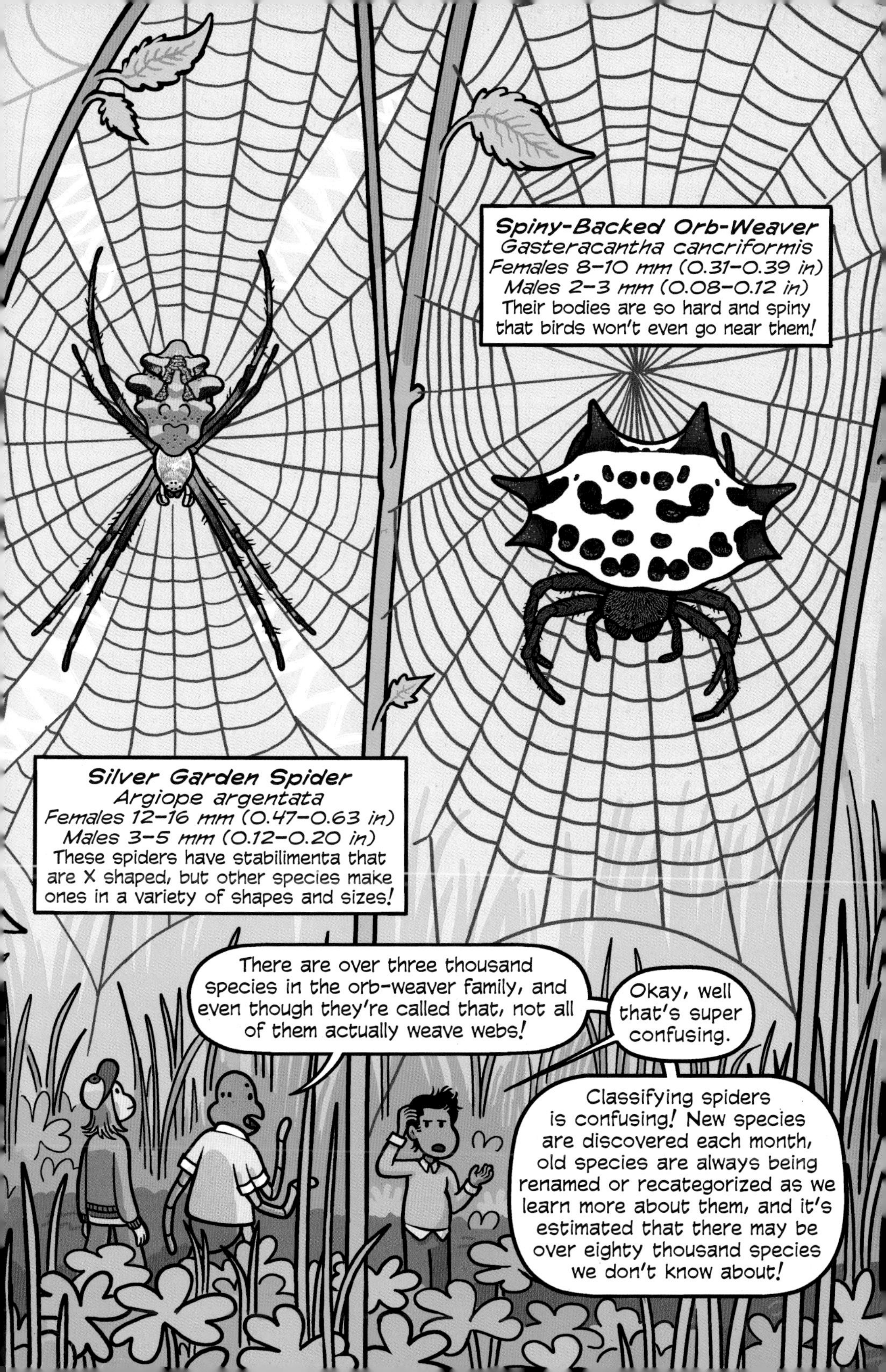
Silver Garden Spider
Argiope argentata
Females 12–16 mm (0.47–0.63 in)
Males 3–5 mm (0.12–0.20 in)
These spiders have stabilimenta that are X shaped, but other species make ones in a variety of shapes and sizes!
Spiny-Backed Orb-Weaver
Gasteracantha cancriformis
Females 8–10 mm (0.31–0.39 in)
Males 2–3 mm (0.08–0.12 in)
Their bodies are so hard and spiny that birds won't even go near them!
There are over three thousand species in the orb-weaver family, and even though they're called that, not all of them actually weave webs!
Okay, well that's super confusing.
Classifying spiders is confusing! New species are discovered each month, old species are always being renamed or recategorized as we learn more about them, and it's estimated that there may be over eighty thousand species we don't know about!

Hmm, let's see if Max is visiting with the ladies in the garden!

Ach! Heads down, ya wee buffoons!

AHHHH

Here's a web-less orb-weaver now! The bolas spider, *Mastophora cornigera*, swings a thread with a sticky droplet on the end of it in a circle—

SWISH SWISH SWISH

HURL

—and hurls it at a passing moth to pull it out of the air!

They also secrete chemicals that mimic the *pheromones** of a moth, which attracts their prey into striking range!

Just like grandma when she goes out dancing!

CHOMP

Yeesh... Yeah, that's Nana all right.

*Scent intended to attract potential mates

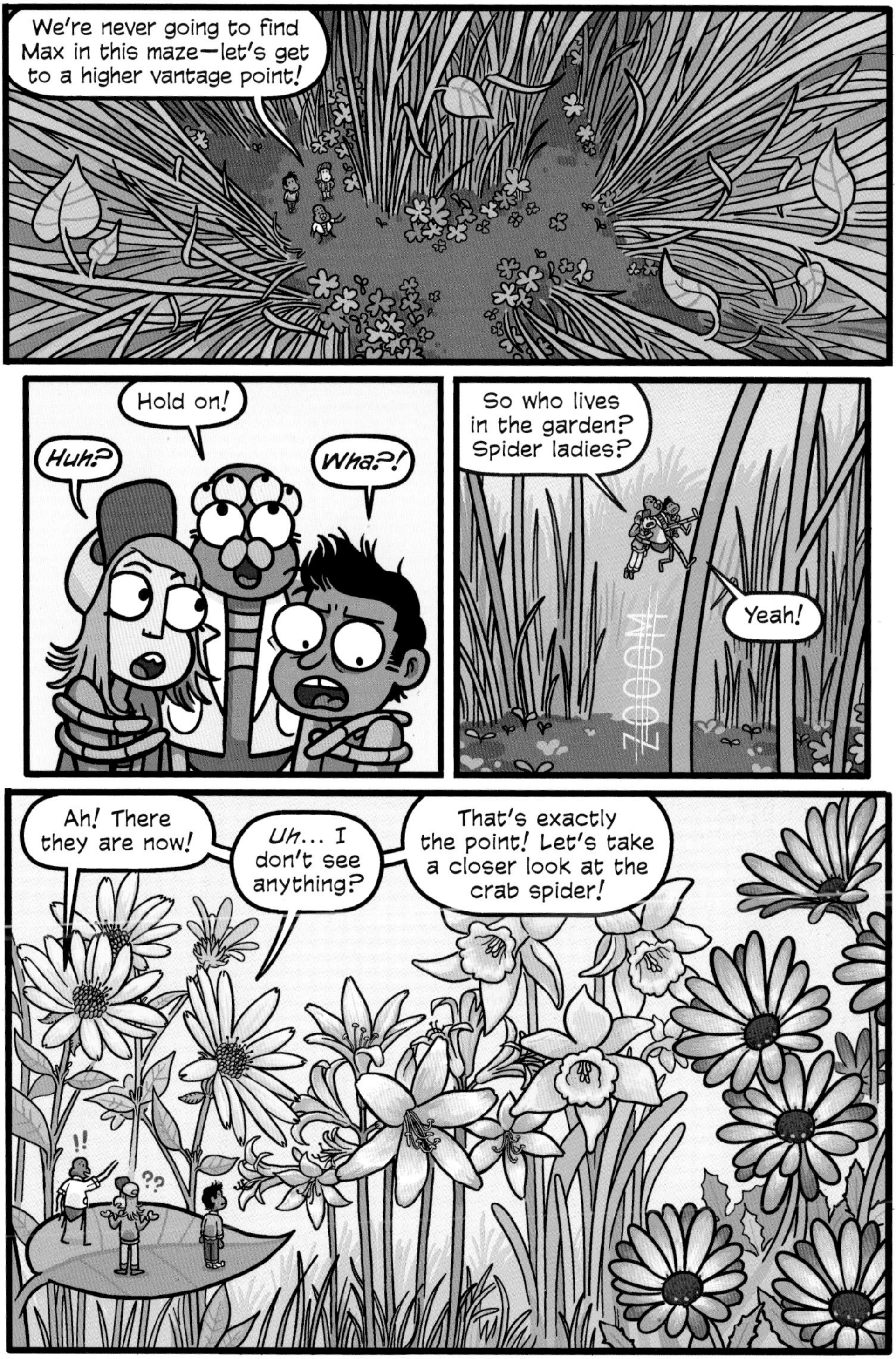
We're never going to find Max in this maze—let's get to a higher vantage point!
Huh?
Hold on!
Wha?!
So who lives in the garden? Spider ladies?
Yeah!
ZOOOM
Ah! There they are now!
Uh... I don't see anything?
That's exactly the point! Let's take a closer look at the crab spider!
!!
??

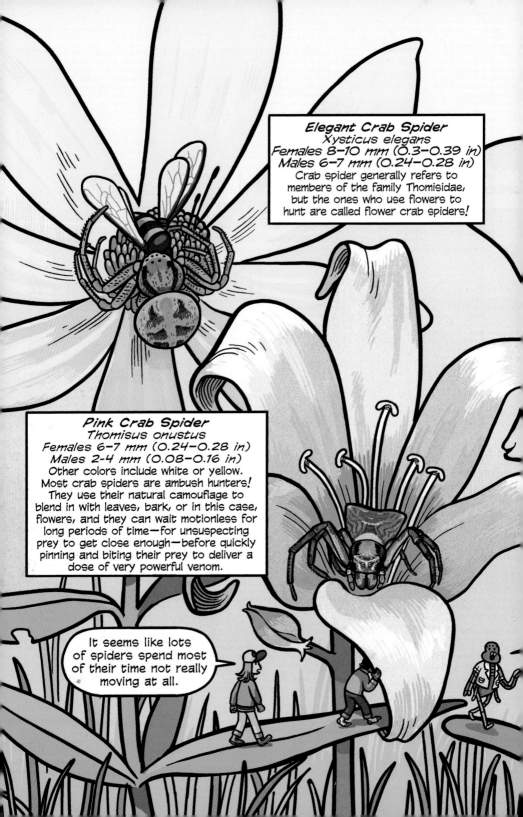

Elegant Crab Spider
Xysticus elegans
Females 8–10 mm (0.3–0.39 in)
Males 6–7 mm (0.24–0.28 in)
Crab spider generally refers to
members of the family Thomisidae,
but the ones who use flowers to
hunt are called flower crab spiders!

Pink Crab Spider
Thomisus onustus
Females 6–7 mm (0.24–0.28 in)
Males 2–4 mm (0.08–0.16 in)
Other colors include white or yellow.
Most crab spiders are ambush hunters!
They use their natural camouflage to
blend in with leaves, bark, or in this case,
flowers, and they can wait motionless for
long periods of time—for unsuspecting
prey to get close enough—before quickly
pinning and biting their prey to deliver a
dose of very powerful venom.

It seems like lots
of spiders spend most
of their time not really
moving at all.

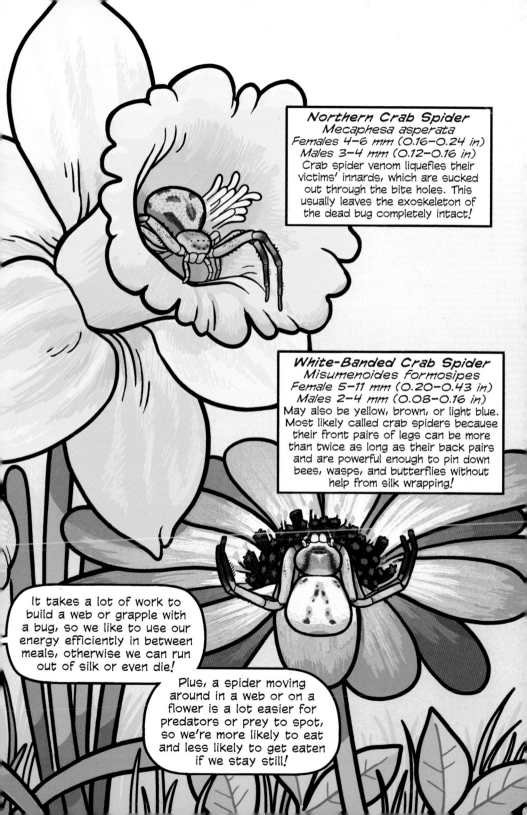

Northern Crab Spider
Mecaphesa asperata
Females 4–6 mm (0.16–0.24 in)
Males 3–4 mm (0.12–0.16 in)
Crab spider venom liquefies their victims' innards, which are sucked out through the bite holes. This usually leaves the exoskeleton of the dead bug completely intact!

White-Banded Crab Spider
Misumenoides formosipes
Female 5–11 mm (0.20–0.43 in)
Males 2–4 mm (0.08–0.16 in)
May also be yellow, brown, or light blue. Most likely called crab spiders because their front pairs of legs can be more than twice as long as their back pairs and are powerful enough to pin down bees, wasps, and butterflies without help from silk wrapping!

It takes a lot of work to build a web or grapple with a bug, so we like to use our energy efficiently in between meals, otherwise we can run out of silk or even die!

Plus, a spider moving around in a web or on a flower is a lot easier for predators or prey to spot, so we're more likely to eat and less likely to get eaten if we stay still!

Misumena vatia, the goldenrod crab spider, can even change color to suit the kind of flower they're on! They hunt on a wide variety of flowers that are varying shades of yellow and white, like daisies and sunflowers.

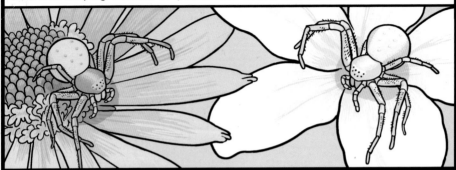

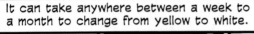
It can take anywhere between a week to a month to change from yellow to white.

Hurry up already!

Slowly, the pigments on its outer layer of cells change to mimic its habitat.

It's possible this is a natural form of crypsis— which means an ability used to avoid detection from potential predators or prey! But it's also possible that they change their color in relation to the amount of sunlight they're receiving as a form of shielding against UV radiation!

Whoa, so it's like sunscreen that their body makes whenever it needs to? I guess that makes sense if they're hanging out on top of flowers in the sun all day.

I don't see Maxie around here, do you?

Couldn't we just ask some of the crab spiders if they've seen them?

Best not to disturb them while they're hunting. We don't want to be mistaken for something tasty!

THWIP

It's really not like them to wander this far from home. If they're not by the stream over there...

What if they wandered into the woods? It's dangerous to be out there all alone!

Um, yeah, we don't have eight legs or cool claws or whatever. Can you help us out here?

Whoops, sorry!

WOBBLE
WOBBLE
WOBBLE

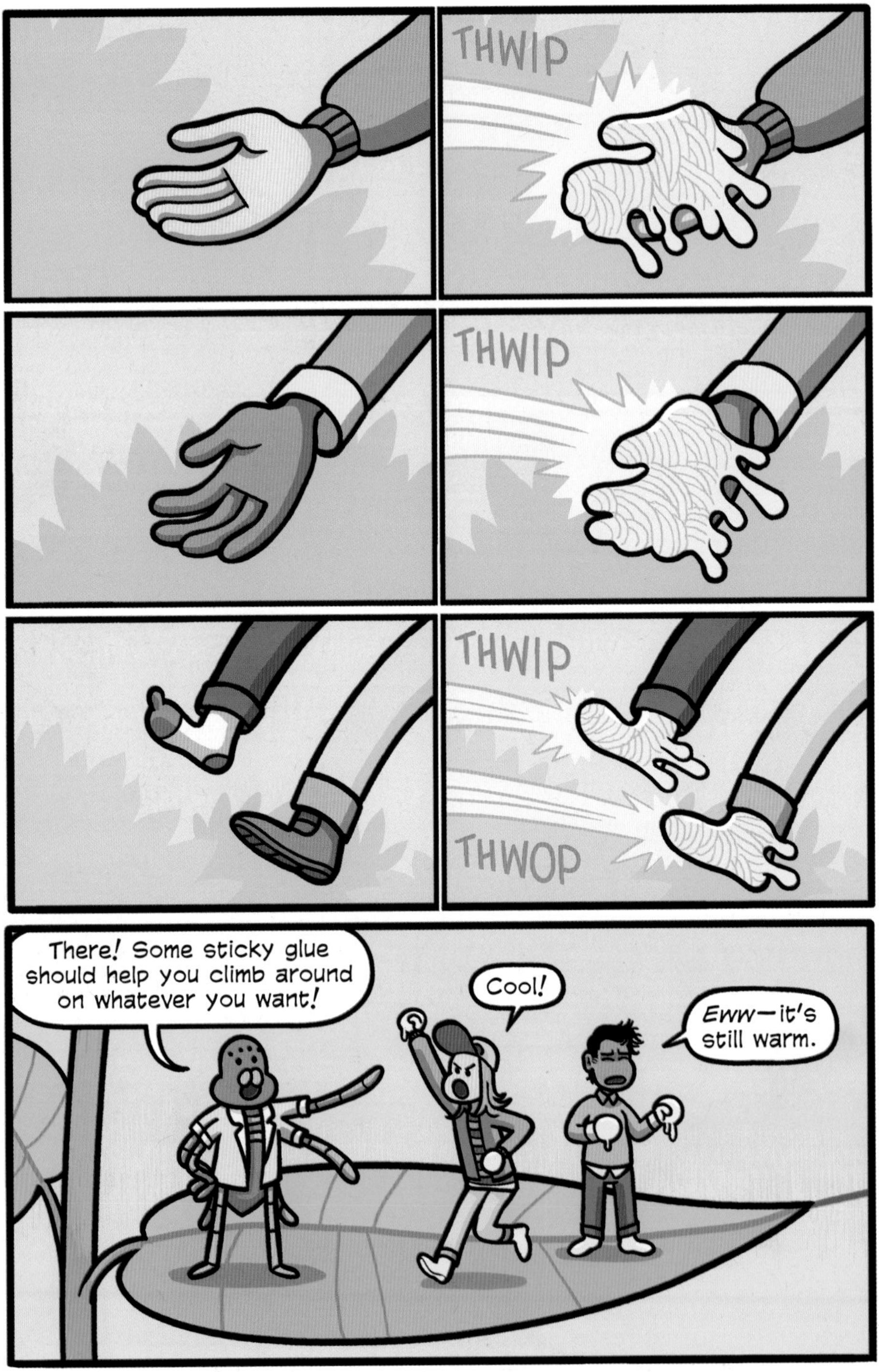

THWIP
THWIP
THWIP
THWOP
There! Some sticky glue should help you climb around on whatever you want!
Cool!
Eww—it's still warm.

Hey—I've noticed something. It seems like in most spider species the males are a lot smaller than the females. Why is that?

Most spiders exhibit sexual dimorphism, which means that males and females have different physical traits from each other! There are lots of animals that are sexually dimorphic aside from spiders too!

Size dimorphism is one of the most common ways in which spiders differ, as the females of most species are much larger than the males. Female *Argiope aemula* can be more than ten times the size of their male counterparts!

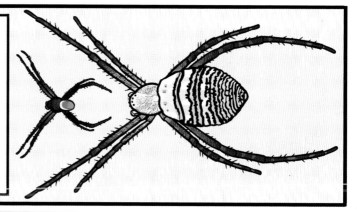

Coloration and body shape is another way to tell males and females apart!

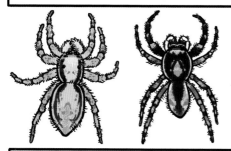

Grey Wall Jumper males and females are the same size but have different patterns on their back.

Spiny-Backed Orb-Weaver males lack red spikes around their abdomen.

Male spiders mature faster too! To grow larger, young spiders molt often.

Beneath their exoskeleton, they grow a larger one, which is softer and folded.

They suspend themselves upside down and shed their old exoskeleton by pushing against it until it cracks.

POP

Then, they slowly wriggle their way free and let their new exoskeleton unfold and harden.

At last! Free from this prison of flesh!

Males are often smaller, so they need fewer molts to become fully mature and reproduce. After his final molt, he abandons web building or even hunting until he finds a female to mate with.

gone matin'

Whereas females' pedipalps resemble a short pair of legs, during their last molt, males gain a special bulb on each palp that is used in fertilization.

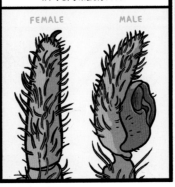

FEMALE   MALE

But before that, he needs to find a female to court! Locating a female that is mature and willing to mate can be difficult.

WEB of LOVE

MATCH NOW

But females secrete pheromones—chemicals that trigger a social or physical response in members of the same species—to attract males. Some leave pheromones on their web or dragline to guide males to them!

When a spider comes across a pheromone web, he eats or destroys it before it can attract other males. Reproduction is the main drive for a mature male, so competition is fierce!

But other guys are the least of his concerns. A female may not be that picky about which things in her web get eaten! Males may wait until a female has fed before trying to mate so that they don't become the meal!

Nooo! I only wanted love!

And it's not just hunger—spiders of all kinds may eat a potential suitor just because they don't want to mate with him! This means spiders have developed some elaborate courtship rituals. This behavior also helps spiders recognize one another as members of the same species when pheromones aren't enough!

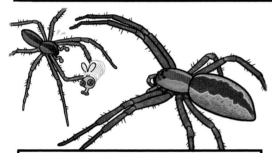

Some male nursery web spiders will catch a fly, wrap it in silk, and bring it to a female as a gift so that she might be less interested in eating him.

A male orb-weaver may attach a thread to a female's web that he plucks rhythmically. Sound and vibration are two important parts of many spiders' courtship rituals!

Some jumping spiders perform complex dances by raising and lowering their abdomen and drumming on the ground.

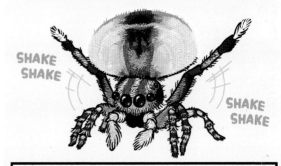

SHAKE SHAKE

SHAKE SHAKE

Maratus volans, the peacock spider, has special flaps on his back. When lifted up, they form a colorful plumage! He waves this along with his legs to court a mate.

Since a male spider's main instinct is to reproduce, he may try several times before running out of energy and dying.

But he may only get one chance! The females of some species, like *Araneus pallidus*, often begin eating the male partway through mating!

CRUNCH

Neat! Isn't that where black widows get their name? I heard they always eat the males as well!

No, that's just their bad reputation! Male black widows usually get away unharmed after mating.

Like most spider species, the male does have a chance of being eaten no matter what, but some *Latrodectus* species even share their web and catch prey with the male for a week or two after mating!

After mating, it takes a few weeks before the female is ready to lay eggs. Most are only about 1 mm (0.04 in) wide, but the number of spiderlings different species give birth to can vary widely!

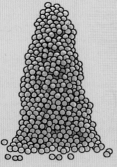

Araneus spiders lay about 1,000 eggs in around ten minutes. A spider's heart rate triples while laying eggs!

But *Cupiennius* have them beat: they can lay up to 2,500 eggs in eight minutes!

Most spiders lay large broods of eggs, but not all! *Oonops* species lay only 2 eggs at a time!

And the tiny *Monoblemma* spiders lay only 1 large egg at a time!

Once the eggs are laid, the spider builds a cocoon of silk around them to protect them from harsh environments or egg eaters like wasps and flies. It also acts as insulation against dangerous temperatures!

The long-bodied cellar spider ties its eggs in a few strands of silk and then carries her egg sac around in her chelicerae until they hatch!

But most egg casings are much more complex! Many spiders, like *Araneus quadratus*, begin by laying out a thin silk disc built up of tightly woven threads.

They climb onto the bottom of the disc and begin laying silk around the edges of it, which slowly builds up into a chamber where they lay the eggs. This process takes about two hours.

Once the eggs are laid, she holds them in with a fine layer of silk and then surrounds the eggs with a thick liquid that hardens and cements them all together.

The spider then covers the structure with a loose thread that she layers up in a mesh pattern. This becomes the hard shell surrounding the entire egg chamber.

I hope we stay this close forever!

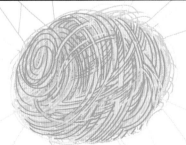

This outer shell may harden or be tightly wound enough that the threads merge.

But there are a lot of ways to wrap our eggs. Let's start with my species!

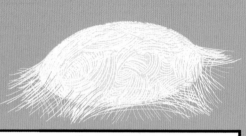

American House Spider
Parasteatoda tepidariorum
Resembles a tiny brown paper bag, but **DO NOT** pack your lunch in it. Trust me on this one.

Grass Spider
Agelenopsis pennsylvanica
A thin contact-lens-shaped web attaches the eggs to leaves or bark. Bad substitute for a normal contact lens unless you like looking at spiders.

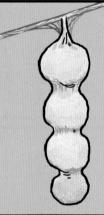

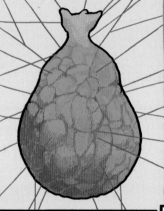

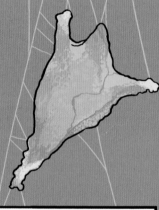

Basilica Orb-Weaver
Mecynogea lemniscata
Multiple egg sacs stuck together with web, like a string of pearls filled with baby spiders!

Yellow Garden Spider
A tough papery shell that looks like a pear but tastes like your worst nightmares.

Silver Garden Spider
Argiope argentata
A flat blob that, let's be honest with each other, looks a lot like a booger.

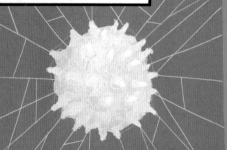

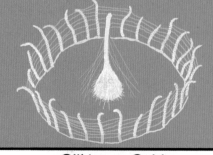

Brown Widow
Latrodectus geometricus
Looks like a fuzzy little explosion, more so when the spiders pop out.

Silkhenge Spider
These incredible egg sacs found in the Amazon are built by a spider that is still unknown to arachnologists! No joke!

Agroeca brunnea creates a small paper-lantern-like structure that includes a molting chamber for the emerging spiderlings to spread into immediately after hatching.

MOLTING CHAMBER

Some spiders—like the jumping spider, *Marpissa rumpfi*—build a nest of layered silk and eggs attached to the ceiling of their burrow.

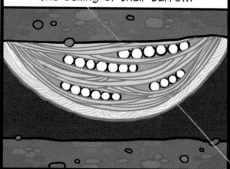

And wolf spiders, members of the family Lycosidae, build very tough round egg casings!

Wolf spiders? Yeah, I think this is where the tour ends for me.

But because they hunt on the go and often don't build webs, they prefer to carry egg sacs around on their back by attaching them to their spinnerets!

In most cases, when the spiderlings are ready to hatch, they use digestive secretions to dissolve the inside layer of web and then gradually push aside the tougher outer fibers to make an opening.

But a wolf spider helps her young get out by biting through the egg sac herself. Without her, the kids aren't able to escape on their own!

CHEW CHEW

I'm finally bustin' outta this joint!

Nyahahahaha! Sweet freedom!

ZOOOOOOOOM

SMAK!

AHH! HELP!

!

—and then it looked at me with its big creepy eyes, and then—

Shhh, don't worry, sweetie. I won't let that mean ol' bird hurt you.

Hey, check it out! Looks like the baby needs his mommy!

Dude...we literally live on her.

SOB SOB

Wolf spiders make great moms. They carry their offspring on their back in a big pile of a hundred or so! The spiderlings live there for the first week of their lives, eating their egg sac until they are big enough to fend for themselves.

Nursery web spiders carry their eggs in their chelicerae until just before the eggs hatch. They then build a tentlike web and hang the egg sac inside of it. The spiderlings stay in the web for a few days once they hatch.

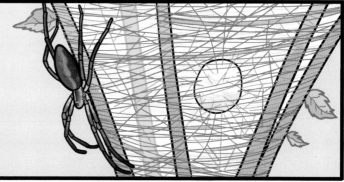

Hey, kids! Dinnertime!

BARF

About twenty species of spiders provide food for their offspring by catching prey for them, regurgitating already liquefied food, or even by laying a batch of unfertilized eggs for the young ones to eat!

Some spiders sacrifice themselves to feed their kids! Matriphagy, or "mom eating," occurs in a few spiders, such as *Cheiracanthium japonicum*, who allow their young to eat them alive, helping them gain weight quicker and giving them a better chance of survival.

YES, EAT OF MY FLESH THAT YOU MAY LIVE!

Num num, sorry, mommy!

CHEW CHEW CHEW CHEW CHEW CHEW CHEW CHEW CHEW

Even more bizarrely, the crab spider *Australomisidia ergandros* converts nutrients from food into hemolymph—the spider equivalent of blood—which is then sucked out through bites in the mother's legs until she is unable to move and is finally devoured fully.

Hey, do you think Mom would do that for us?

Ewww, shut up!

OH!!

That's Max's knapsack! They must have passed through here earlier!

Don't worry, I'm sure they dropped it by accident!

I hope so...

DROP

Do you mind carrying this? The creek is just up ahead, and Max may have tried to cross on their own...

Sure!

Here you go!

SHOVE

Don't you worry about all this water ruining your webs?

Not really! Spider silk is fairly waterproof, so wet environments aren't much of a problem for us!

Watery areas where insects like to lay eggs also provide us with tons of food!

Spiders are great at fishing too! There are species from a few families that specialize in hunting on or *in* water!

Is nowhere safe?!

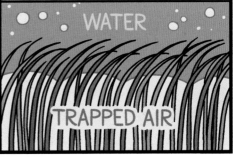

The majority of spiders will drown in a pool or river, but some species have a covering of water-repellant hairs called a hydrophobic cuticle, which keeps them dry, even when they're fully submerged in water!

WATER

TRAPPED AIR

Some water spiders use the wind to sail across the surface of the water.

Others row across the surface with their legs.

When attacking prey, some can even gallop across the top of the water at five times their normal speed!

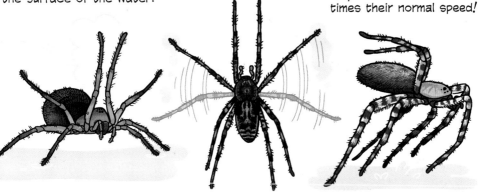

But the water is a dangerous place too! Spiders in open water can make a quick meal for a passing fish or frog.

Oh, here we go! This is a great spot to see some fishing spiders in action!

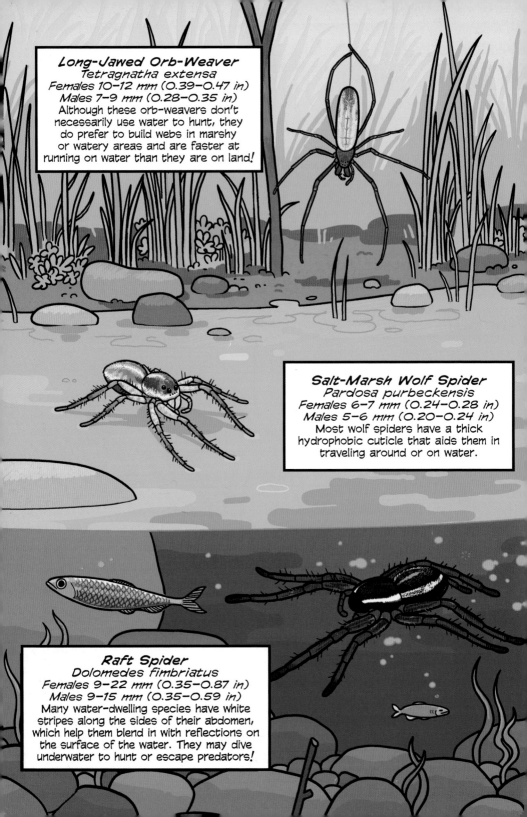

Long-Jawed Orb-Weaver
Tetragnatha extensa
Females 10–12 mm (0.39–0.47 in)
Males 7–9 mm (0.28–0.35 in)
Although these orb-weavers don't
necessarily use water to hunt, they
do prefer to build webs in marshy
or watery areas and are faster at
running on water than they are on land!

Salt-Marsh Wolf Spider
Pardosa purbeckensis
Females 6–7 mm (0.24–0.28 in)
Males 5–6 mm (0.20–0.24 in)
Most wolf spiders have a thick
hydrophobic cuticle that aids them in
traveling around or on water.

Raft Spider
Dolomedes fimbriatus
Females 9–22 mm (0.35–0.87 in)
Males 9–15 mm (0.35–0.59 in)
Many water-dwelling species have white
stripes along the sides of their abdomen,
which help them blend in with reflections on
the surface of the water. They may dive
underwater to hunt or escape predators!

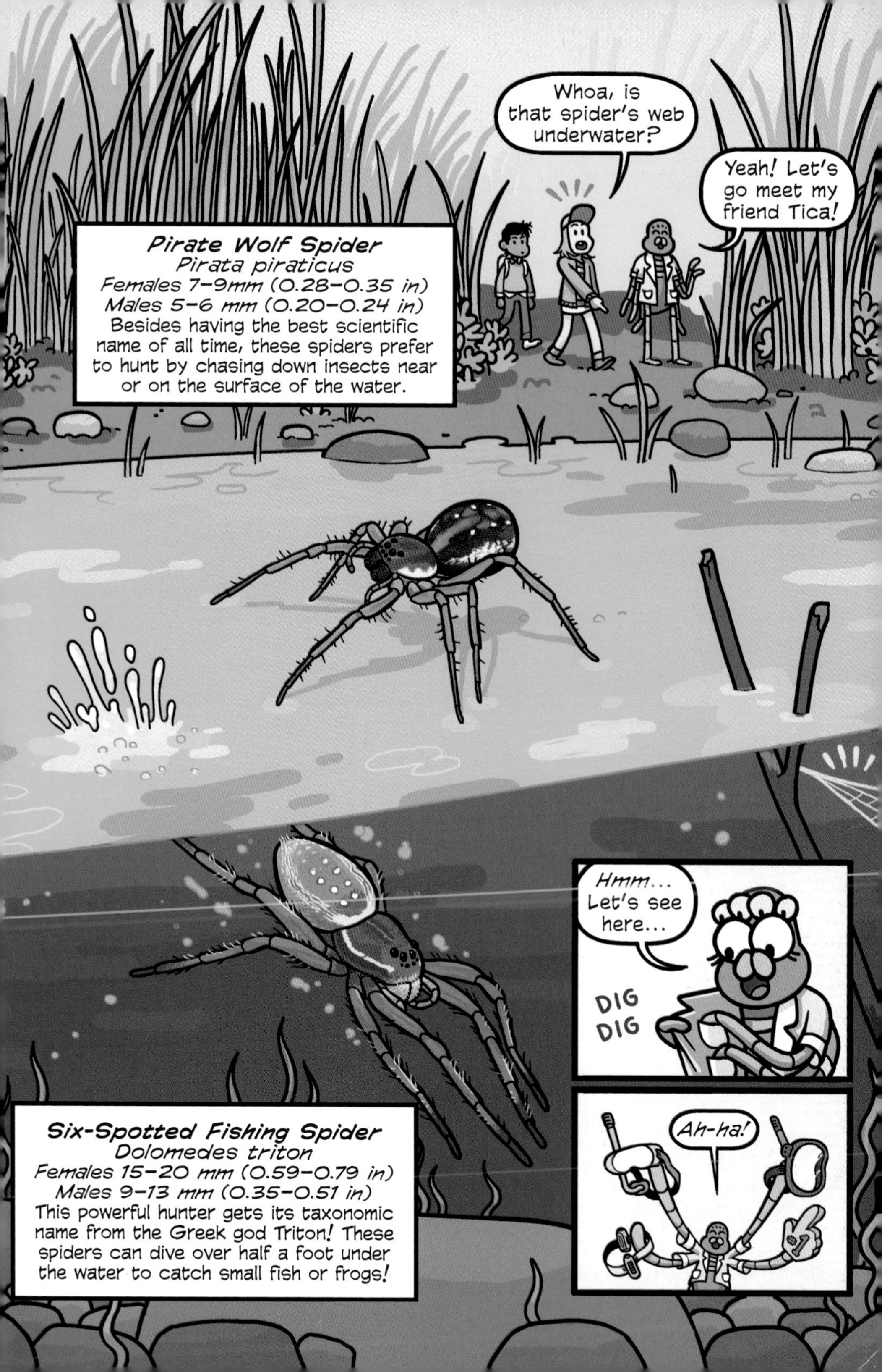
Whoa, is that spider's web underwater?
Yeah! Let's go meet my friend Tica!
Pirate Wolf Spider
Pirata piraticus
Females 7–9mm (0.28–0.35 in)
Males 5–6 mm (0.20–0.24 in)
Besides having the best scientific name of all time, these spiders prefer to hunt by chasing down insects near or on the surface of the water.
Hmm... Let's see here...
DIG DIG
Ah-ha!
Six-Spotted Fishing Spider
Dolomedes triton
Females 15–20 mm (0.59–0.79 in)
Males 9–13 mm (0.35–0.51 in)
This powerful hunter gets its taxonomic name from the Greek god Triton! These spiders can dive over half a foot under the water to catch small fish or frogs!

BLAAA

LAST ONE IN IS A ROTTEN SPIDER EGG!

SPLOOSH

Hey, what's that belt thingy?

Another one of my inventions! It lets me breathe while I'm underwater!

See, spiders don't breathe through our mouths! Small slits on the underside of our abdomen lead to organs called book lungs. These alternating hollow stacks of tissue allow us to filter oxygen out of the air!

Hemolymph, the spider equivalent of blood, passes through alternating stacks and absorbs oxygen to be transported to the spider's other organs and tissues.

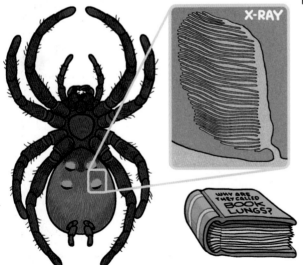

X-RAY

WHY ARE THEY CALLED BOOK LUNGS?

HEMOLYMPH

AIR & OXYGEN

In addition to book lungs, or sometimes instead of them, spiders may have trachea that branch throughout their body. Spiders with a tracheal system (usually smaller, modern spiders) are better at retaining water and less prone to dehydration!

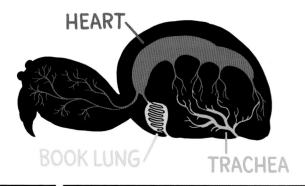

HEART

BOOK LUNG

TRACHEA

Book lungs may have actually evolved from fish gills, as the structure of the two are similar!

WATER

Spiders have what is called an open circulatory system, which means that hemolymph is pumped into the body cavity of the spider by its heart.

I'm just a big bag of blood!

This cavity, the hemocoel, is filled with passageways that hemolymph travels through as it washes over the internal organs to provide them with oxygen.

Oh good, it looks like he's home!

This is my friend the diving bell spider, *Argyroneta aquatica!*

Tepi! Hi, y'all! Wow, it's not often that I get this many visitors at once!

I'd invite you in but it's a little cramped in here!

Is that your home? It's cool!

Oh this old thing? It may not look like much, but it's perfect for little ol' me!

BLOOP

How much time do you spend underwater?

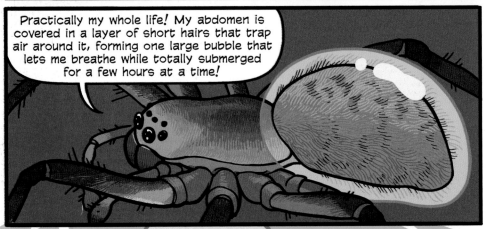

Practically my whole life! My abdomen is covered in a layer of short hairs that trap air around it, forming one large bubble that lets me breathe while totally submerged for a few hours at a time!

The key to living down here is our *diving bell*—

:grab:

CHOMP

For one, it's where we mate and lay eggs.

But it's also where we wait to ambush fish or insects that pass by.

We bring anything we catch back to the bell so that our venom isn't diluted with water!

A diving bell starts with a sheet web, but as we add layers, we also like to spice this baby up a li'l bit!

RUB RUB RUB

We fill the spaces between the threads with a unique protein-rich hydrogel. This creates the outer layer of the bell that allows air to be trapped inside.

We form the bubble we live in by transporting air on our backs from the surface and brushing it off to get trapped underneath the web. Hydrogels behave similarly to tissue like skin in that they are absorbent and made up of mostly water!

FWISH FWISH

OXYGEN

CO2

Because of this, *gas exchange* can take place between the bubble and the water outside. Animals are performing gas exchange when they breathe! Their cells take in oxygen and give off carbon dioxide, and the lungs pump those gases in and out of the body.

Both gases in the bubble exert different amounts of pressure on it, so when the spider breathes in or out it forces the bubble to equalize the pressure by removing or adding gas! The water contains more oxygen than the air in the bubble, so oxygen flows through the hydrogel and replenishes the air inside, while carbon dioxide the spider exhales flows out!

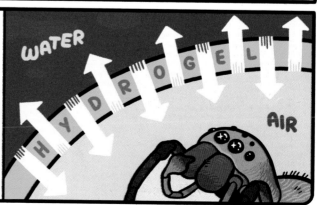

WATER

HYDROGEL

AIR

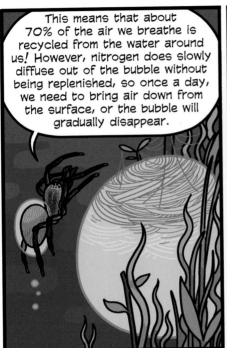

This means that about 70% of the air we breathe is recycled from the water around us! However, nitrogen does slowly diffuse out of the bubble without being replenished, so once a day, we need to bring air down from the surface, or the bubble will gradually disappear.

Ooh, speaking of which, mine could use some replenishing. I assume you're heading wherever Max was going?

You saw Maxie? When was this?!

WHERE?!

WHAT DID THEY SAY?!

Hey, calm down! I didn't talk to them, but they looked fine! It was half an hour ago. They were dragging leaves around up there.

OOOOH, they're heading right into the woods! We'd better go catch up to them. That forest is full of predators!

Thanks for everything!

Good luck!

⋛*Brrrrr*⋚ I should have brought a towel... I didn't realize we were gonna get totally soaked!

You were the one who jumped in the stream!

SHIVER

Oh, hold on just a minute!

SPIN SPIN

Here you go!

Ooh! Thank you, Tepi!

Okay, you two, keep an eye out for Maxie. What could they have been doing with leaves? Look for any leaves that might be a clue!

How are we gonna be able to tell a clue leaf from a regular leaf?

PLOP

Well, you see, if we look at the, *oh, um...*

...the leafy part...

...of the leaf...*here*... we should be able to...

Oooh, it's useless! My vision is awful, I can't tell the difference between any of these stupid things!

TOSS

Wait, really? I thought spiders had great vision! Don't you need to be able to see the bugs and stuff you catch? Or see things that are trying to eat you?

Most spiders don't rely on vision to tell what's going on in the world!

Well then what's with all those freaky eyes?

Eyes are just one of our major sensory organs! You humans rely on them so much you've neglected to evolve the other ways of "seeing" the world around you!

Anyway, most of us have eight eyes, but not all of us! Some have six, four, or two eyes!

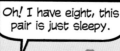

Wait, Tepi, doesn't your species have eight eyes? Why do you only have six?

hm?

Oh! I have eight, this pair is just sleepy.

ZZZ

So don't try anything funny back there!

POP

About 99% of spiders have eight eyes arranged in two or three rows on the front of their face. Each set of eyes has a name to help when comparing spiders! We break them up into four groups to describe which set we're talking about!

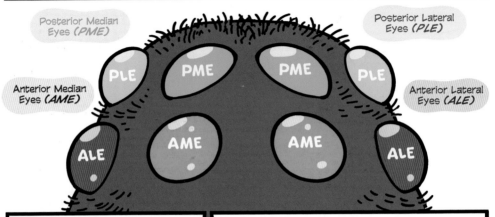

Posterior Median Eyes (*PME*)

Posterior Lateral Eyes (*PLE*)

Anterior Median Eyes (*AME*)

Anterior Lateral Eyes (*ALE*)

The AME are a spider's primary set of eyes. They usually appear black, since they lack an organ called a tapetum.

But this reflective surface in the inner eye is present in all of their secondary eyes. It can make these eyes appear to be different colors because they're so reflective!

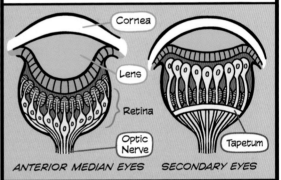

Cornea

Lens

Retina

Optic Nerve

Tapetum

ANTERIOR MEDIAN EYES   *SECONDARY EYES*

Their reflectiveness also makes them more sensitive to light. In wolf spiders, the PME and PLE contain 15–35 times the light-sensitive cells of the AME!

Eye arrangement is helpful for arachnologists and arachnophiles to identify unknown species because many within a family or genus share similar faces!

Ant-Mimic Crab Spider
Amyciaea lineatipes

Wolf Spider
Alopecosa fabrilis

Magnolia Green Jumper
Lyssomanes viridis

Mangrove Jumper
Ligurra latidens

Goldenrod Crab Spider
Misumena vatia

Trap-Jaw Spider
Chilarchaea quellon

Southern Black Widow
Latrodectus mactans

Trapdoor Spider
Liphistius desultor

Wood-Louse Hunter
Dysdera crocata

Brown Recluse
Loxosceles reclusa

Malagasy Green Lynx Spider
Peucetia madagascariensis

Cobalt Blue Tarantula
Cyriopagopus lividus

Most members of the cribellate spider family Caponiidae have two eyes, but a few members of their family have four, six, or eight as well!

And a few species have no eyes at all! The eyeless huntsman spider, *Sinopoda scurion*, doesn't need them, since it lives deep in caves where there's no light at all! It relies entirely on its other sensory organs to hunt and survive.

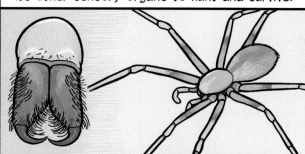

And don't forget jumping spiders! These big-eyed babies are the largest spider family by far, making up 13% of all spiders in the world!

Yes! My favorite! No offense, Tepi...

None taken!

Those gorgeous colors! Those furry li'l palps! Those big ol' eyes!

THEY ARE THE CUTEST SPID—

doo dee doo

AHHH!

WOOSH

EEK!

Ah! Pardon me, I had no intention of frightening you!

Zebra Jumper
Salticus scenicus
Females 5-9 mm (0.20-0.35 in)
Males 5-6 mm (0.20-0.24 in)

Ohhhh! Even their labels are cute!

CRUNCH CRUNCH

JUST LOOK AT THOSE EYES! Tepi! Get me the shrink ray and set it on reverse. I'm taking a sweet furry angel home with me!

Ahem, excuse me, madam, but my eyes are *HIGHLY* evolved organs!

Jumping spiders' AME have highly developed retinas. The eye is elongated with many extra layers of cells called photoreceptors.

that's these things

These cells respond to light and send visual signals to our brains, which interpret what we see!

bzzzzz

They spot prey with the other sets of eyes, which are better at detecting movement, and then quickly orient themselves toward their prey to focus with their AME before jumping.

They're much better for spotting prey than say, an orb-weaver's! Some spiders have fine vision at close range, but ours is better at spotting prey that's farther away!

And they're so cuuute!

Stop that!

Spiders without big adorable baby eyes still have several other powerful senses!

Some of the longer hairs that cover a spider's body contain sensitive nerves that allow them to sense touch and vibrations on the ground.

DOINK

Hey, quit pokin' me!

Although their legs and bodies are covered in thousands of these hairs, touching just one is enough to trigger a fight-or-flight response!

We also have specialized hairs in lines along our legs called trichobothria! These amazing organs can sense even the tiniest vibrations in the air and are specifically tuned to pick up the vibrations of insect wings beating.

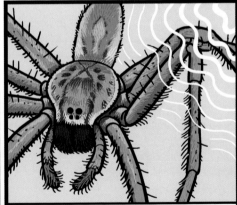
Cupiennius spiders can detect a fly that's 30 centimeters (11.81 in) away, from just the vibration of its wings!

Spiders also have small slits covering their bodies and legs that sense bends and pressure on their exoskeleton.

This is vital because their outer layer is a hard shell that can't be deformed past a certain point without breaking!

They also sense vibration! Pirate spiders use them to locate prey struggling in the water nearby!

Yarr! Avast!

SPLISH SPLASH

Spiders also "taste" with their legs and palps using specialized hairs. They can tell the difference between a rotting and freshly dead bug and identify harmful chemicals or poisons they should avoid.

Eww! Who left this in here?!

Hey, make yourself useful! Jump up real high and look for a spider kid around here!

Fine, if you promise to leave me alone forever.

When jumping spiders take off, they are traveling at about 85 cm per second (1.9 miles per hour) and they can go up to 16 cm (6.3 in) in the air. For a human, that would be like jumping to the top of a nine-story building!

They're behind you.

SHIFF

WHEEEE

Oh, hey, Mom!

MAX! GET BACK HERE!

Meet me down-river at the big treeeeeeee...

Where did this recklessness come from?!

Well, that actually looks pretty fun...

And they're steering pretty well with just the sail!

Wow, they must've used piriform discs to attach the aft—

WAIT!

THERE'S A BEND AHEAD IN THE RIVER! THEY'RE GOING TOO FAST!

Okay, gonna need you to get up real high and see where the nearest huntsman spider is!

I can JUMP, you know...

WOOOSH

So? Which way do we go?

SLAM!

SOUTHWEST! NOW LET ME SUCK THE GUTS OUT OF THIS BUG IN PEACE! I'M LUCKY I WASN'T KILLED!

That's not luck, that's physics, baby! When falling, small things reach their *terminal velocity*, or maximum speed, faster than larger things. So a spider falling from the top of a skyscraper will land at roughly the speed it left the roof at, without being harmed, but if a whale falls, its speed *multiplies* until it hits the ground.

BANK CORP. LTD

Come on! The huntsman is this way!

It sort of sounds like they, *um*...hunt man?

Oh, don't worry. They're called that because they are solitary hunters who chase down or ambush prey. They're afraid of humans!

I mean, okay, but we're pretty small, so, like—

We need one because they're fast runners!

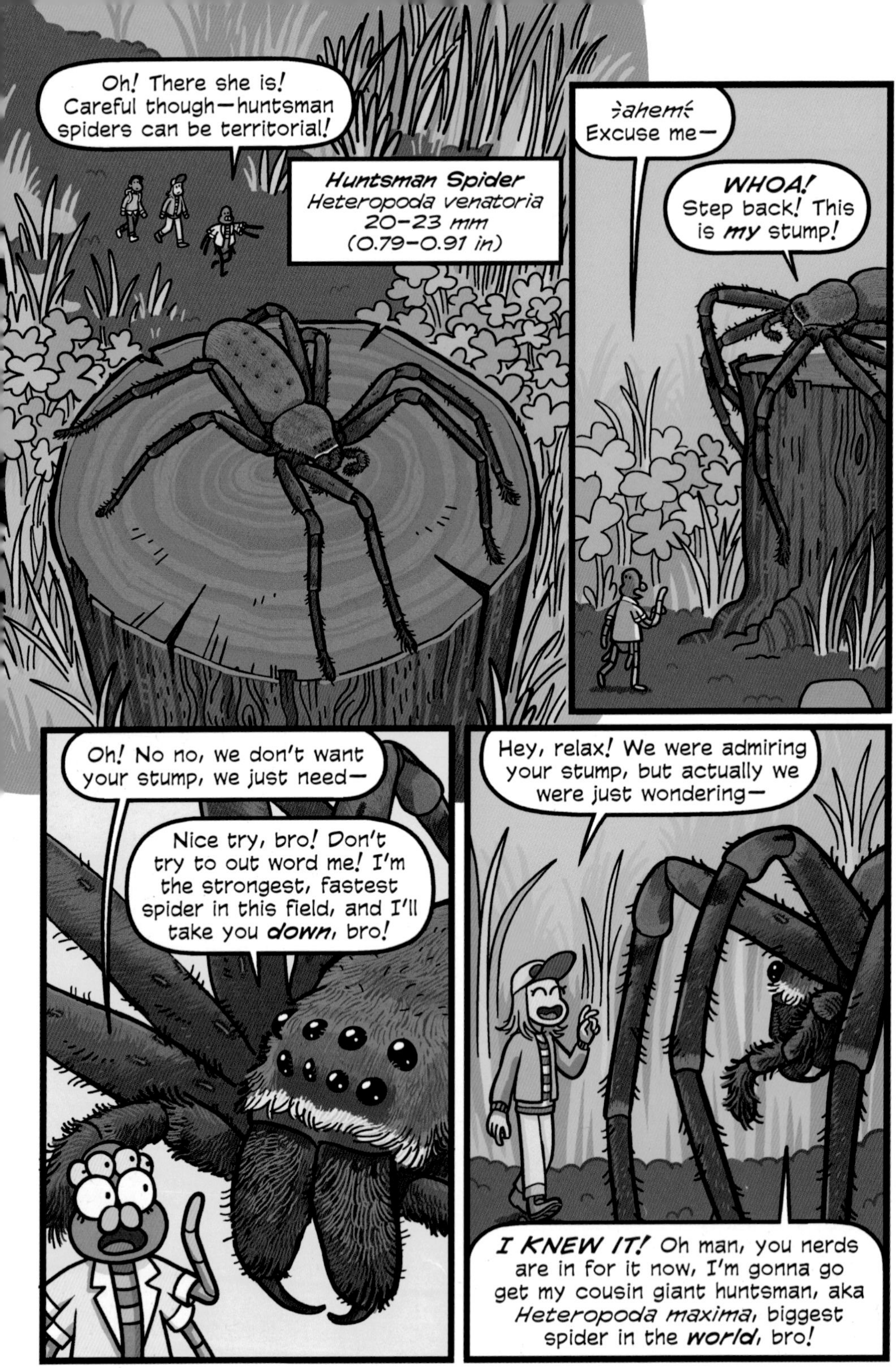
Oh! There she is! Careful though—huntsman spiders can be territorial!
Huntsman Spider
Heteropoda venatoria
20–23 mm
(0.79–0.91 in)
≯ahem≮ Excuse me—
WHOA! Step back! This is my stump!
Oh! No no, we don't want your stump, we just need—
Nice try, bro! Don't try to out word me! I'm the strongest, fastest spider in this field, and I'll take you down, bro!
Hey, relax! We were admiring your stump, but actually we were just wondering—
I KNEW IT! Oh man, you nerds are in for it now, I'm gonna go get my cousin giant huntsman, aka Heteropoda maxima, biggest spider in the world, bro!

Well...biggest in the world by leg span! Not the largest in size, *that* would be—

DON'T CARE! I'm the biggest in the field; I'm the fastest. Everyone, shut up!

Actually, the real reason we're here is because we heard from the other bugs around here that you're *not* the fastest spider in this field...bro...

WHAT?! Who told you that? Was it Kevin? That little worm is in for it!

Wait! I've got an idea: why don't you just show us how fast you run?!

BRO! That's perfect!

And we'll all ride on your back to increase the challenge! That'll show Kevin!

BRO-BRO! That's perfecter!

All right, grab some hair!

Eep!

75

The way spiders move looks complex, but really, it's a simple back-and-forth movement of two pairs of legs at once!

They move their first and third pairs of legs at the same time, and then their second and fourth, stepping one pair after another.

In most animals, arm and leg movement is controlled by two kinds of muscles—flexors and extensors. Flexor muscles contract and extensors...well, extend!

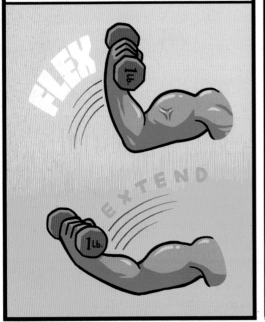

But spiders don't have extensors in their legs! They have only flexors for pulling their legs back inward! To extend them, they rapidly pump hemolymph into their legs with their back muscles, using hydraulic blood pressure to push their legs out!

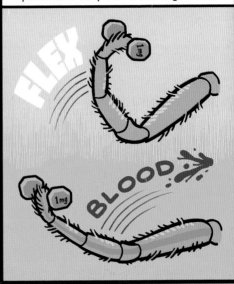

One and three touch the ground while two and four are lifted off the ground and vice versa!

When the spider wants to turn, it takes longer strides on either the left or right side, causing it to move in the opposite direction!

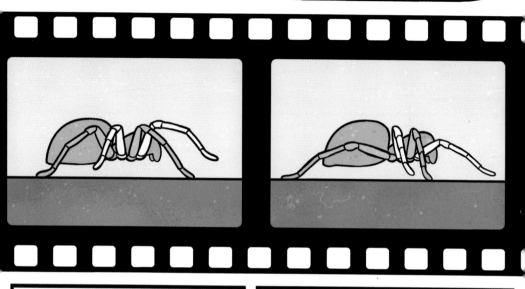

This is why spiders curl up into a ball when they die: as their heart slows, they lose blood pressure and aren't able to push their legs out. A dehydrated spider may go through this same process!

Alas! I am no more!

Spiders also have an incredible defense mechanism built into their legs! When threatened or trapped, they can detach a leg at will to escape!

So long, dingus!

Shedding a piece of the body when threatened is called *autotomy!* This defense mechanism is shared by some other animals, like lizards, crabs, and even mice!

It isn't even necessary for a predator to be pulling on the limb for it to come off! And it's totally up to the spider when the leg goes. Spiders under anesthesia aren't able to perform this process!

POP

When the leg is fully detached, the muscles surrounding the joint close naturally to prevent a fatal loss of blood pressure. But as long as the spider hasn't gone through its final molt yet, it doesn't need to worry!

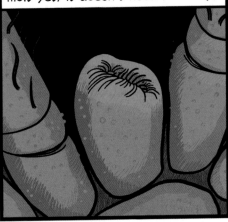

That's because spiders can actually regenerate their lost limbs!

If they lose a limb in the first half of the period between molts, it will start to grow back slowly!

This leg grows curled up inside the stump of the old leg, so it may grow to be thinner or smaller than the others.

BEFORE MOLTING

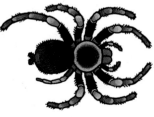

AFTER MOLTING

AFTER LAST MOLT

Is this the fastest a spider can go?

Nope, the fastest spider is the giant house spider, *Eratigena atrica*, who moves at almost 1.9 kilometers per hour (1.2 mph)!

FWOOOOOOSH

But some spiders forgo walking for more interesting transportation methods!

The Moroccan flic-flac spider, *Cebrennus rechenbergi*, does leaping aerial front flips to escape from predators.

BOING

Wheeee!

The wheel spider, *Carparachne aureoflava*, rolls down sand dunes on its side by curling its legs and flexing its joints to propel itself!

Hey, *uh*, how are we gonna stop when we need to?

The dragline acts as an anchor, so a spider just clamps down its spinnerets to come to a quick stop!

MAX'S BOAT!

Hey, you can stop now! You've proved you're the fastest!

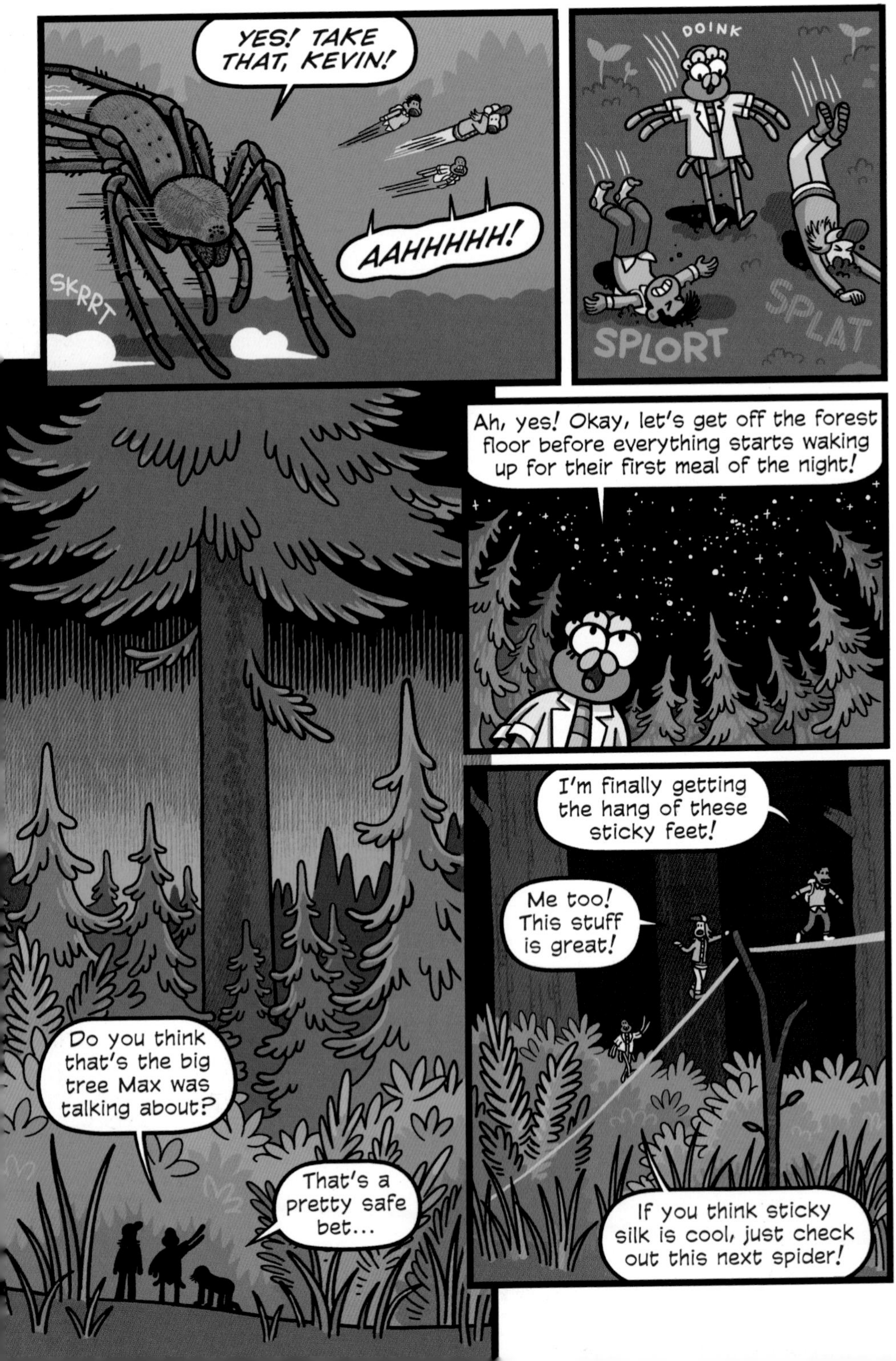
YES! TAKE THAT, KEVIN!
AAHHHHH!
SKRRT
DOINK
SPLORT
SPLAT
Ah, yes! Okay, let's get off the forest floor before everything starts waking up for their first meal of the night!
Do you think that's the big tree Max was talking about?
That's a pretty safe bet...
I'm finally getting the hang of these sticky feet!
Me too! This stuff is great!
If you think sticky silk is cool, just check out this next spider!

Some spiders are *diurnal*: they hunt and build webs during the day. Others are nocturnal, or active at night! This one is a night hunter with a specialized silk that helps it quickly subdue prey.

Spitting Spider
Scytodes thoracica
Female 4–6 mm (0.16–0.24 in)
Male 3–4 mm (0.12–0.16 in)
These long-range hunters unleash a specialized silk from their cheliceral opening. This silk comes from their venom glands, making a potent mix of both silk proteins and venom!

It spits envenomed sticky silk in a zigzag, covering its prey in less than one seven-hundredth of a second.

FSSSH

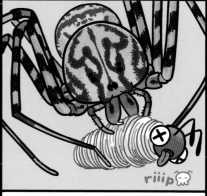

The silk immobilizes the prey so that the spider can move in and wrap it with regular silk before delivering a fatal bite.

riiip

They're also *scavengers*: they'll happily eat a dead insect or other animal if they come across it.

If spider eyes are so bad, how do they see at night?

Our vision is terrible, but our eyes are highly light sensitive!

Spiders' sleeping and eating cycles are mostly determined by the time of day due to the changes in available light.

Ahh, another beautiful day for eating bugs!

Do you remember the reflective layer in some spiders' eyes—the tapetum? At night, it reflects light off the back of the eye, increasing the amount of light the eye can use, similar to the way night-vision goggles work!

This makes it easy to locate some species, like wolf spiders, at night! If you hold a flashlight level with your eyes and point it into a field, you may see one or several pairs of tiny green eyes flashing back at you!

That's probably a wolf spider! Holding the light at eye level will reflect it right back at your face. Otherwise, it's easy to miss!

Ogre-face spiders, members of the genus *Deinopis*, have night vision that is twice as powerful as a cat's or owl's!

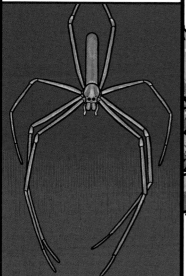

Although these spiders have eight eyes, their PME are extremely enlarged and face forward, making them look particularly menacing. These spiders' eyes lack tapeta. Instead, they produce a super-light-sensitive membrane every single night that breaks down in the morning!

GRRR!

These cribellate spiders hold a small web between their front two pairs of legs, held straight out as they hang vertically off the ground. They spot prey with their powerful eyes and drop the net on it as it passes by.

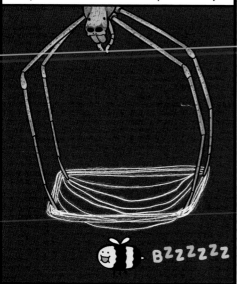

BZZZZZZ

They remain motionless while hanging vertically, camouflaging themselves against the branches of the trees where they live!

Oh! Look over there!

California Trapdoor Spider
Bothriocyrtum californicum
20–32 mm (0.79–1.26 in)

Quelle surprise!

Trapdoor spiders live in silk-lined underground burrows. After the burrow has been dug, they make a silk disc, which they cover with dirt and dry brush.

The disc is attached with thick hinges of silk, making the trapdoor that covers the entrance to their home. They grip the inside with their chelicerae to keep it shut.

Some lay lines of silk leading out from their burrows and wait for the vibrations of an insect passing across them.

Some will jump out and chase down passing prey.

ZOOOOOOM

Some barely leave their burrow at all, only attacking things within easy reach.

yoink!

This helps protect them from other predators as well! Although many kinds of spiders also live in a burrow, the trapdoor is a unique advantage for this family of spiders.

Some trapdoor spiders, such as *Cyclocosmia* and *Galeosoma*, **are** the trapdoor! These spiders have a distinct abdomen with an intricate "seal" on the bottom that provides camouflage in soil.

PECK!!!
PECK
PECK

When threatened, they retreat headfirst into their burrow, where their tapered, flat back creates a plug that prevents predators from getting to their softer bits!

Wouldn't living in a hole in the ground get lonely? Spiders are so solitary!

There's a small group, called social spiders, who spend loads of time together!

"Social spiders" isn't a group like "wolf spiders"—which is the informal term used to describe members of the family Lycosidae. Social spiders are what we call any species that cooperates in prey capture and/or brood care.

Velvet spiders, members of the family Eresidae

The huntsman spider *Delena cancerides*

Lynx spiders from the genus *Tapinillus*

Communal hunting is one of the great advantages social spiders have over others, as multiple spiders can work to take down larger prey than one spider could on their own. A bug caught in the communal web of *Anelosimus eximius* is shared by the whole group, meaning that any individual spider in the colony can go longer without catching prey and still survive.

Whoa! There are a lot of you here. Are we having a party or something?

The colony also provides a wider selection of mating partners, help caring for spiderlings, and better defense against predators!

Anelosimus eximius also builds some of the largest webs in the world! Communal webs can reach lengths of 7.62 meters (25 feet) and may contain upward of fifty thousand spiders. These huge webs can blanket whole trees or even forests when colonies join together.

Oh no, I think this is all my fault...

What do you mean?

Okay, this *has* to be the tree.

Max. I know why they came to *this* tree...

It's an ant colony! I've been researching ant mimics, spiders who have mastered the art of disguise! Max must've come to do research!

We'd better put on some disguises if we're going to try to enter an ant colony! Those things do *not* like outsiders!

Disguises?

That's right! And we'll make them ourselves.

Ants have been a viable food source for about sixty million years, and a variety of spiders have evolved over that time to better hunt them by copying their appearance, behavior, and even the chemicals they secrete!

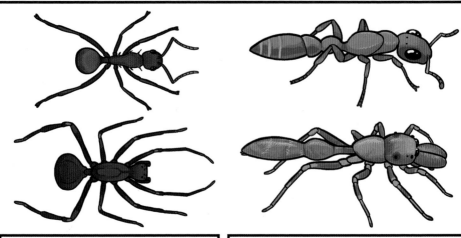

Their body may be constricted, making it appear that they have three body segments, like an ant, and their legs are thinner than usual.

Some have enlarged spots on their head that look like an ant's eyes, and their enlarged chelicerae mimic an ant carrying something in its mandibles!

Okay, let's see your best ant faces!

Wow, we look absolutely terrible.

Yeah, I don't know if I'm gonna fool any ants with this disguise.

Oh, don't worry, here's the most important part!

ANT PHEROMONES QUEEN♥SIZE

FSSSH

≥COUGH COUGH≥

≥COUGH≥

Sorry, but it's necessary! Ants leave a trail of pheromones wherever they go, for other ants to follow, so you need to be dripping with this stuff.

≷HACK≷

spritz spritz

When we join the line of ants, try to act like one of them, and don't sweat! It'll wash the pheromone right off you!

Uh... I'll try?

To blend in, spiders have to move like ants too. They can walk on their back three pairs of legs while holding up their front pair to mimic an ant's antennae!

They also move their abdomen up and down as they walk, like an ant does, making it easier to move about inside a colony without alerting the ants to their presence.

wiggle wiggle

Mimicry helps spiders stay alive too! Ants are a lot less palatable than a nice juicy spider, plus attacking an ant means a predator may have to deal with the whole colony, so blending in may save a spider from becoming the next meal!

The weaver ant mimic has long chelicerae that resemble an ant's head! Spiders may also mimic ant movements to blend in or to lure them away from the pack!

It's also common for ant mimics to walk in long winding circles like an ant following a pheromone trail!

Look, it's Maxie's hat! They must've gone this way!

Fascinating! So you spend your whole life here?

Yeah, now get outta here, kid, you're blowin' my cover!

MAX!

Oh, hey, Mom! You'll never believe what I found! It's—

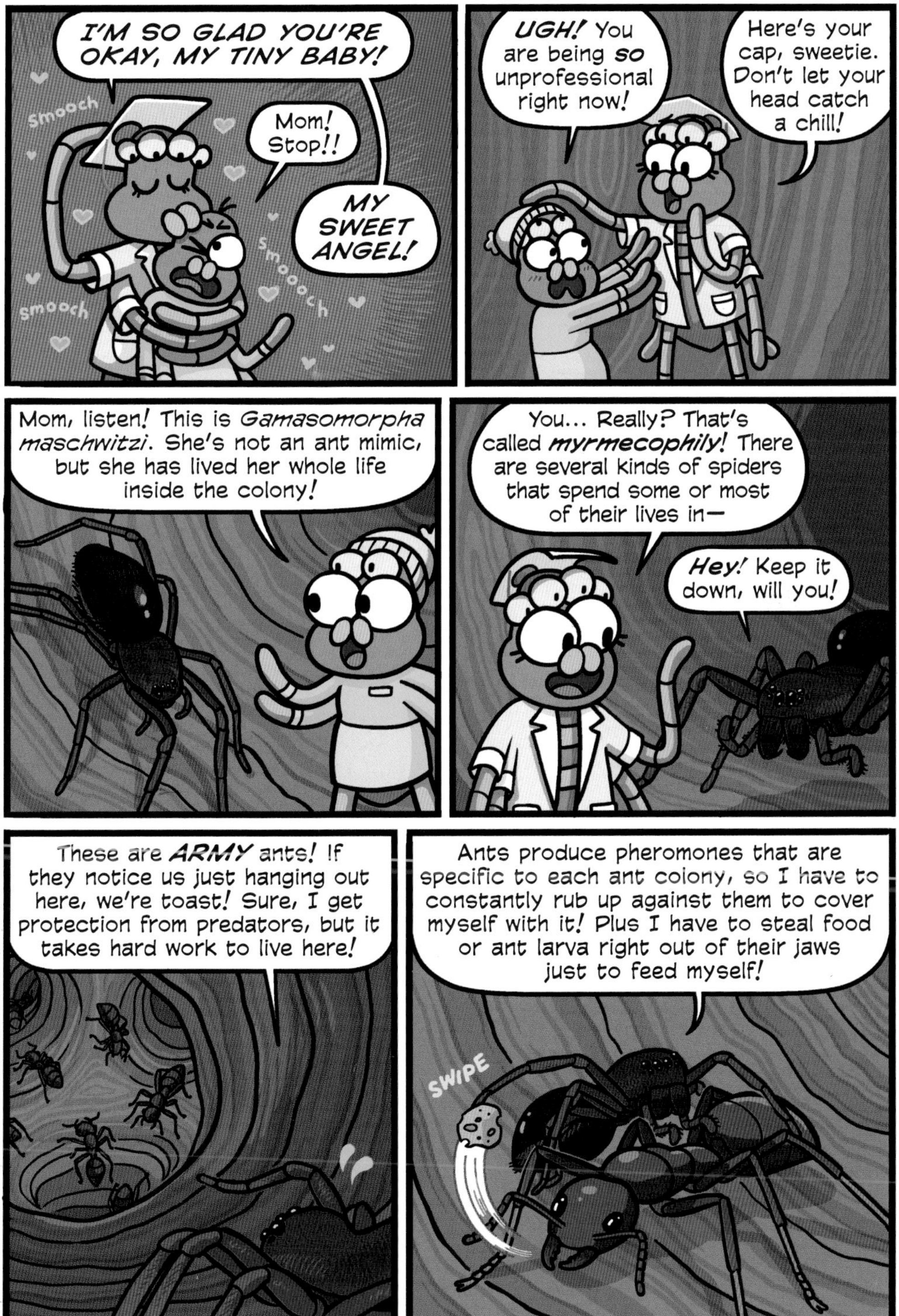
I'M SO GLAD YOU'RE OKAY, MY TINY BABY!
smooch
Mom! Stop!!
MY SWEET ANGEL!
smoooch
smooch
UGH! You are being so unprofessional right now!
Here's your cap, sweetie. Don't let your head catch a chill!
Mom, listen! This is Gamasomorpha maschwitzi. She's not an ant mimic, but she has lived her whole life inside the colony!
You... Really? That's called myrmecophily! There are several kinds of spiders that spend some or most of their lives in—
Hey! Keep it down, will you!
These are ARMY ants! If they notice us just hanging out here, we're toast! Sure, I get protection from predators, but it takes hard work to live here!
Ants produce pheromones that are specific to each ant colony, so I have to constantly rub up against them to cover myself with it! Plus I have to steal food or ant larva right out of their jaws just to feed myself!
SWIPE

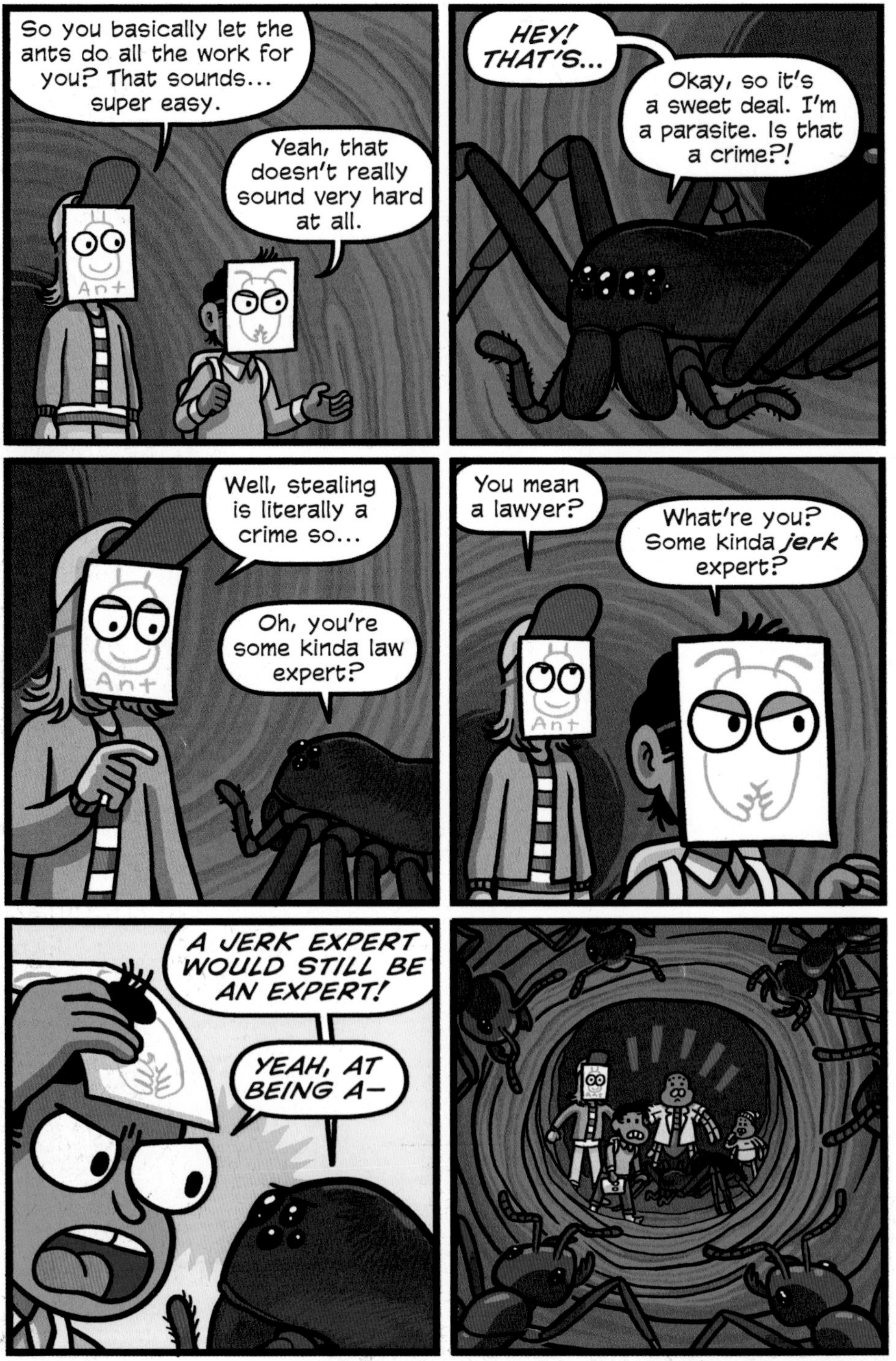
So you basically let the ants do all the work for you? That sounds... super easy.
Yeah, that doesn't really sound very hard at all.
Ant
HEY! THAT'S...
Okay, so it's a sweet deal. I'm a parasite. Is that a crime?!
Well, stealing is literally a crime so...
Oh, you're some kinda law expert?
Ant
You mean a lawyer?
What're you? Some kinda *jerk* expert?
Ant
A JERK EXPERT WOULD STILL BE AN EXPERT!
YEAH, AT BEING A—

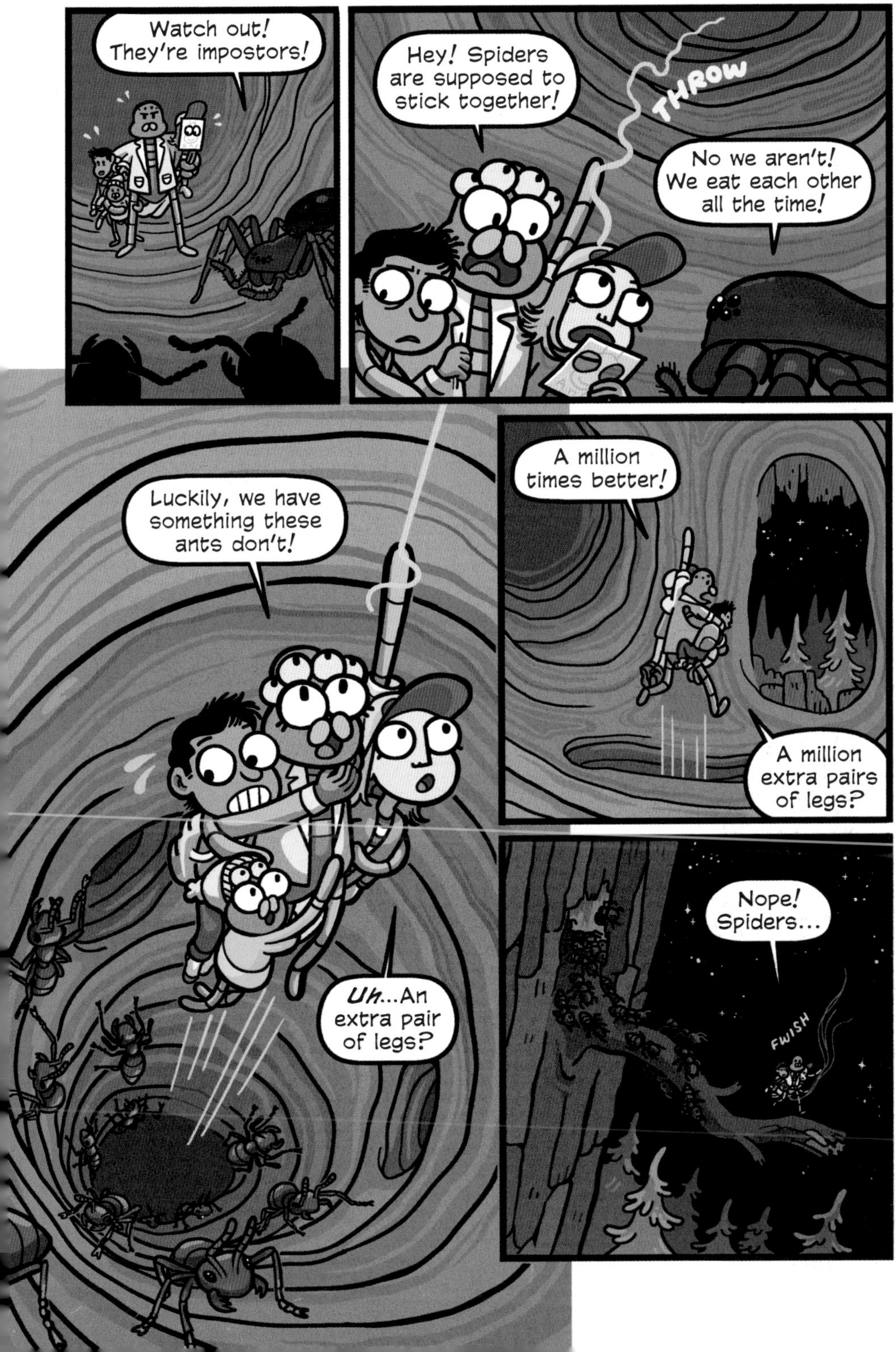
Watch out! They're impostors!
Hey! Spiders are supposed to stick together!
THROW
No we aren't! We eat each other all the time!
Luckily, we have something these ants don't!
Uh...An extra pair of legs?
A million times better!
A million extra pairs of legs?
Nope! Spiders...
FWISH

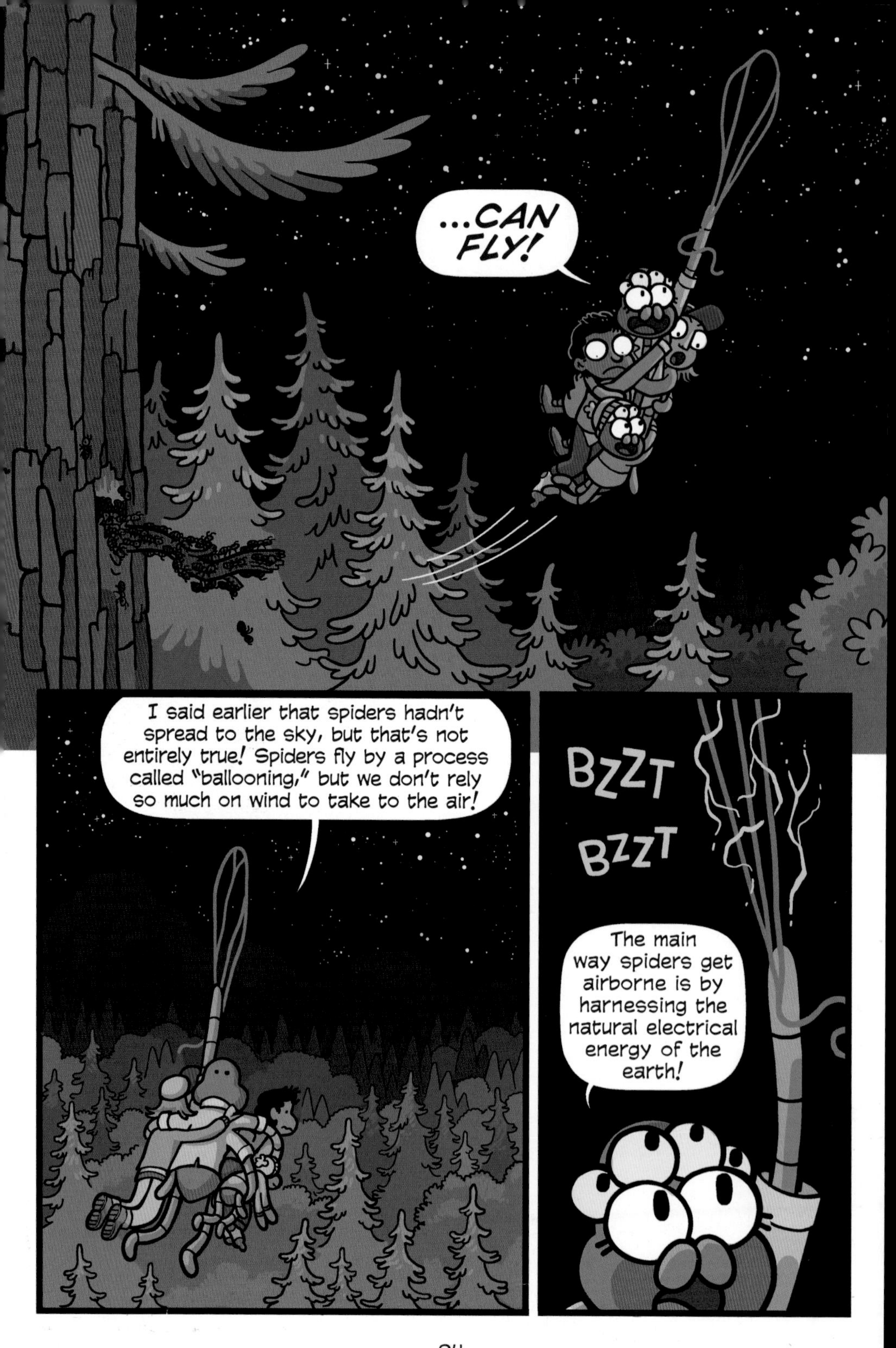
...CAN FLY!
I said earlier that spiders hadn't spread to the sky, but that's not entirely true! Spiders fly by a process called "ballooning," but we don't rely so much on wind to take to the air!
BZZT
BZZT
The main way spiders get airborne is by harnessing the natural electrical energy of the earth!

In 1901, a scientist named Nikola Tesla began building an enormous device on Long Island in New York. Wardenclyffe Tower, as it was called, was conceived of as an experimental wireless-message transmission station.

But he soon theorized it could also transmit power wirelessly! He knew that the sun's upper atmosphere constantly releases solar wind—a stream of protons, electrons, and other electrically charged particles.

This stream passes continuously over the earth, so many of the particles accumulate in our atmosphere. The negative charge of the earth attracts the positive charge in the atmosphere, leading to a higher density of particles and a higher electrical charge closer to the earth's surface!

Tesla thought this meant he could turn the entire earth into a conductor and wirelessly transmit power to homes and factories worldwide. He was never able to test this, as all of his investors pulled out before the tower was completed.

Can I have two million dollars so I can give free electricity to everyone on earth forever?

Ummm... No.

NIKY'S FREE ELECTRICITY TOWER

BIG BUCK'S POWER CO.

BUCK

But what Tesla didn't know was that spiders had already harnessed that same electrical field millions of years ago to conquer the planet.

A study done in a windless chamber showed that spiders are still able to balloon if the chamber is charged with a mild electrical field. It also showed that their trichobothria hairs are *electroreceptive*, meaning they can detect changes in electrical fields!

Part of why webs are so good at capturing insects is because spider silk usually carries a negative charge. When a positively charged insect hits the web, static electricity pulls the silk toward the bug!

This positively sucks!

Don't be so negative!

Before ballooning, a spider finds the highest point possible and prepares by sticking its abdomen in the air and standing on its claws, decreasing the amount of contact with the ground.

When it detects a powerful enough positive charge in the air with its trichobothria, it releases a few threads of negatively charged silk.

THWIP

FWOOSH

If the charge in the air is powerful enough, the "kite" gets pulled off the ground, and the spider along with it!

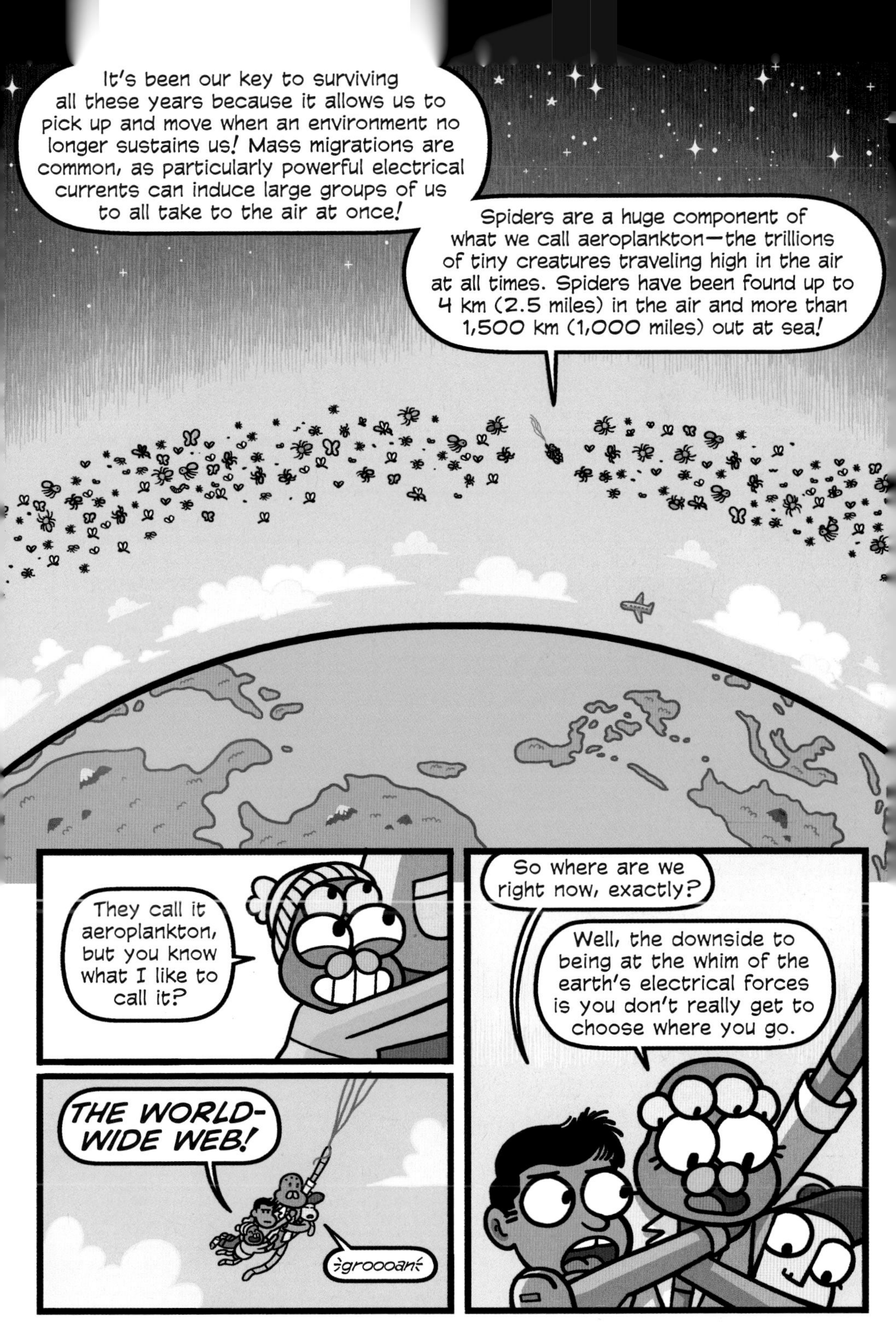

It's been our key to surviving all these years because it allows us to pick up and move when an environment no longer sustains us! Mass migrations are common, as particularly powerful electrical currents can induce large groups of us to all take to the air at once!
Spiders are a huge component of what we call aeroplankton—the trillions of tiny creatures traveling high in the air at all times. Spiders have been found up to 4 km (2.5 miles) in the air and more than 1,500 km (1,000 miles) out at sea!
They call it aeroplankton, but you know what I like to call it?
THE WORLD-WIDE WEB!
groooan
So where are we right now, exactly?
Well, the downside to being at the whim of the earth's electrical forces is you don't really get to choose where you go.

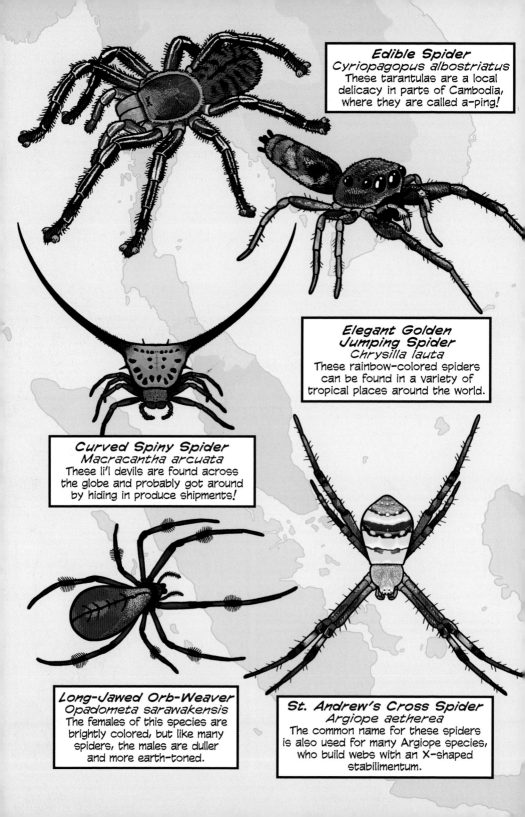

Edible Spider
Cyriopagopus albostriatus
These tarantulas are a local
delicacy in parts of Cambodia,
where they are called a-ping!

**Elegant Golden
Jumping Spider**
Chrysilla lauta
These rainbow-colored spiders
can be found in a variety of
tropical places around the world.

Curved Spiny Spider
Macracantha arcuata
These li'l devils are found across
the globe and probably got around
by hiding in produce shipments!

Long-Jawed Orb-Weaver
Opadometa sarawakensis
The females of this species are
brightly colored, but like many
spiders, the males are duller
and more earth-toned.

St. Andrew's Cross Spider
Argiope aetherea
The common name for these spiders
is also used for many Argiope species,
who build webs with an X-shaped
stabilimentum.

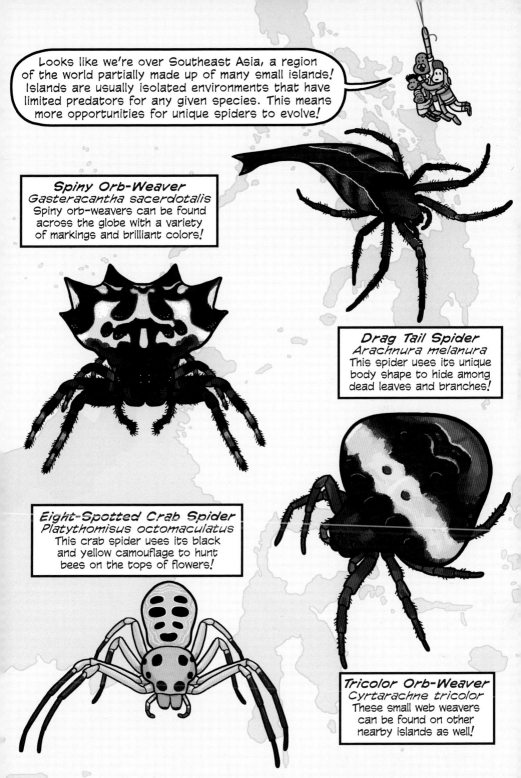

Looks like we're over Southeast Asia, a region of the world partially made up of many small islands! Islands are usually isolated environments that have limited predators for any given species. This means more opportunities for unique spiders to evolve!

Spiny Orb-Weaver
Gasteracantha sacerdotalis
Spiny orb-weavers can be found across the globe with a variety of markings and brilliant colors!

Drag Tail Spider
Arachnura melanura
This spider uses its unique body shape to hide among dead leaves and branches!

Eight-Spotted Crab Spider
Platythomisus octomaculatus
This crab spider uses its black and yellow camouflage to hunt bees on the tops of flowers!

Tricolor Orb-Weaver
Cyrtarachne tricolor
These small web weavers can be found on other nearby islands as well!

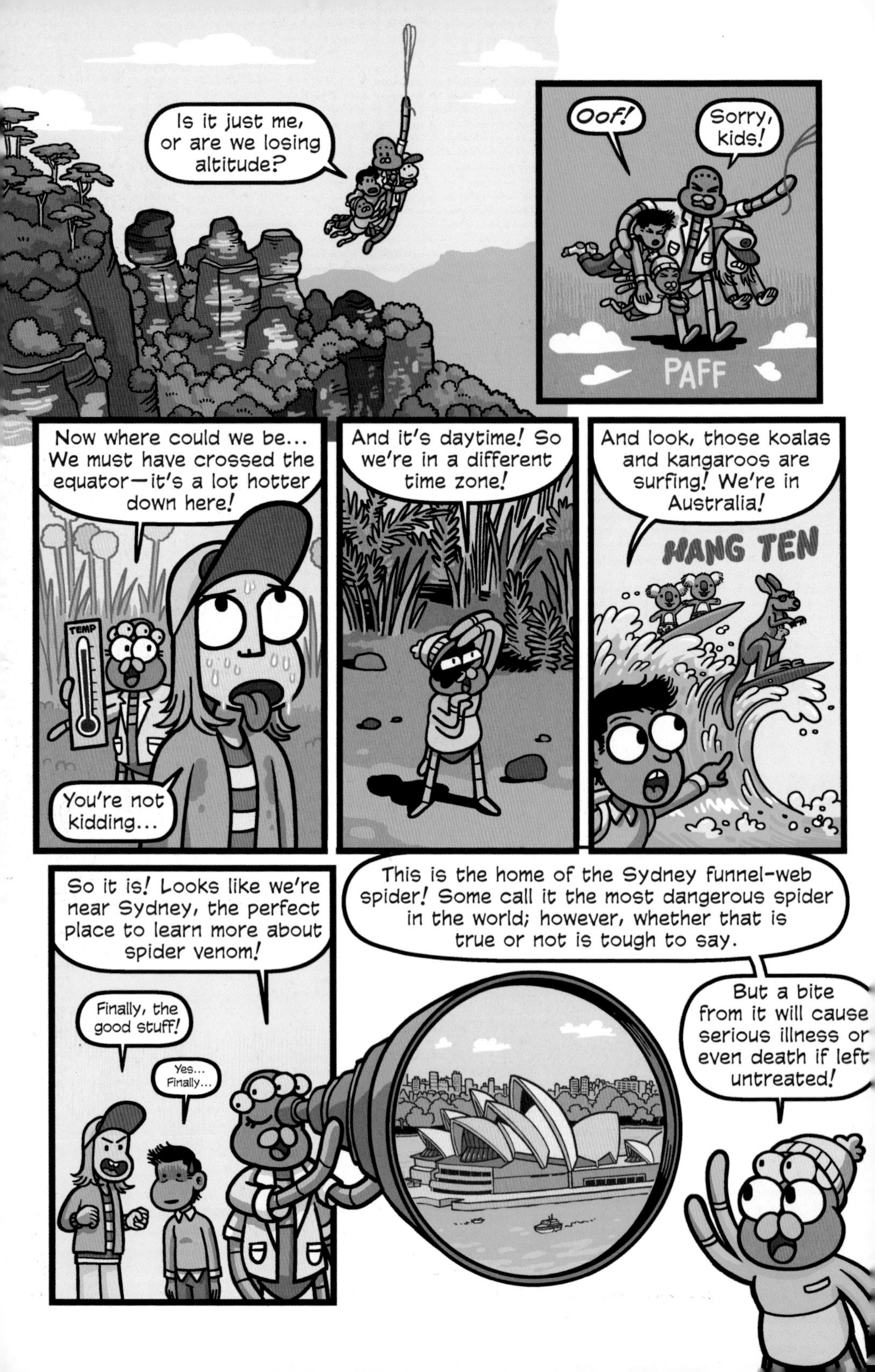
Is it just me, or are we losing altitude?
Oof!
Sorry, kids!
PAFF
Now where could we be... We must have crossed the equator—it's a lot hotter down here!
TEMP
You're not kidding...
And it's daytime! So we're in a different time zone!
And look, those koalas and kangaroos are surfing! We're in Australia!
HANG TEN
So it is! Looks like we're near Sydney, the perfect place to learn more about spider venom!
Finally, the good stuff!
Yes... Finally...
This is the home of the Sydney funnel-web spider! Some call it the most dangerous spider in the world; however, whether that is true or not is tough to say.
But a bite from it will cause serious illness or even death if left untreated!

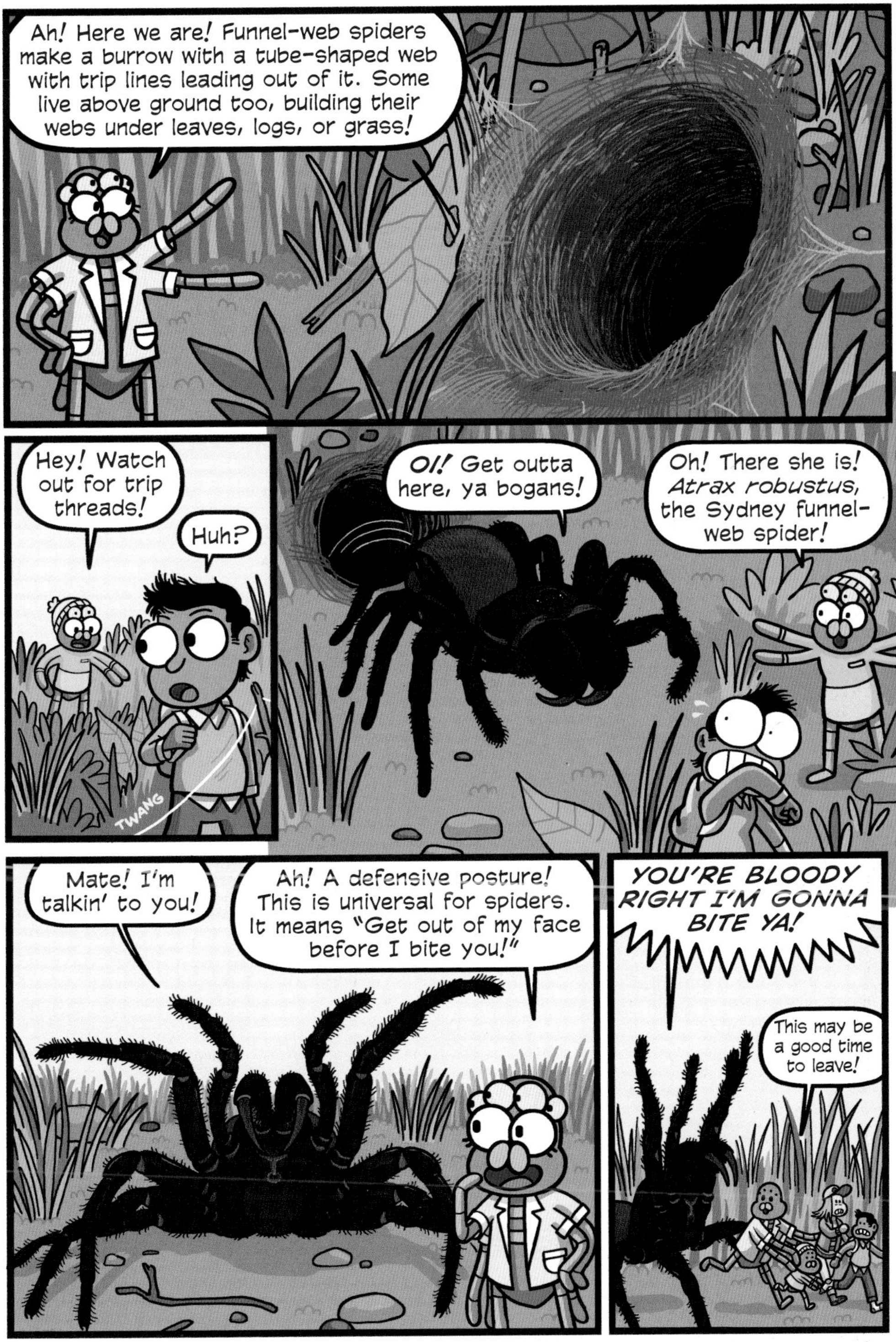

Ah! Here we are! Funnel-web spiders make a burrow with a tube-shaped web with trip lines leading out of it. Some live above ground too, building their webs under leaves, logs, or grass!
Hey! Watch out for trip threads!
Huh?
TWANG
OI! Get outta here, ya bogans!
Oh! There she is! *Atrax robustus*, the Sydney funnel-web spider!
Mate! I'm talkin' to you!
Ah! A defensive posture! This is universal for spiders. It means "Get out of my face before I bite you!"
YOU'RE BLOODY RIGHT I'M GONNA BITE YA!
This may be a good time to leave!

Aww, it was just gettin' good!

Max!

COME GET US, UGLY!

That's it! If there's one thing I can't tolerate, it's rude comments about the appearance of others! Get ready for the biting of your life!

Max! You know better than that... Although if one of us gets bitten, it would present a fascinating opportunity to collect data!

WHAT?! I'M NOT GETTING BITTEN FOR DATA!

Oh fine, but you're missing out! Venom is made up of a bunch of stuff like amino acids and enzymes, which create specific reactions in the venom, or between the venom and bodily fluids.

Cytotoxins attack tissue. They can liquefy an insect's innards, but in large animals, they cause blisters around the bite as well as necrosis, aka tissue death. This is typical of bites from spiders like the brown recluse.

Neurotoxins shut down or impair the central nervous system, so when a human gets a dose of a neurotoxin-heavy venom like that of the Sydney funnel-web spider, it causes an overproduction of neurotransmitters, which causes paralysis and death.

SYMPTOMS of S.F.W.S. BITE

sweating
muscle
spasms
watery
eyes
tongue
twitching
drooling

Hey, soooo now seems like a good time to tell us what to do if we get bitten.

Oh right! Well, first, avoid bugging spiders at all!

Not every spider's venom will make you sick, but that doesn't mean you shouldn't be careful around them, especially if you don't know what kind of spider it is!

The first thing to do if you get bitten is tell an adult! For most bites, cleaning the wound with soap and water and using an antibiotic ointment is enough. Ice can help reduce swelling and pain.

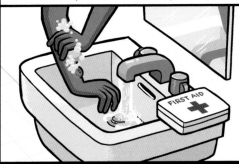

FIRST AID

But if the bite causes extreme or lasting pain, muscle cramps, fever, nausea, or loss of breath **OR** if you think the bite might be from a dangerous spider, you should seek medical help immediately!

GASP

Fortunately, for almost every dangerous spider on the planet, humans have made an antivenom! A tiny bit of a specific kind of venom is injected into an animal like a sheep or goat, and their immune system creates antibodies to fight it.

Now hold on just a sec...

These antibodies are most of what makes up an antivenom. They bond with the toxic venom components in your body and neutralize them.

ANTIBODY

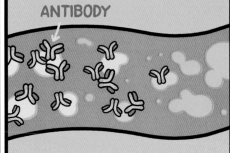

This system is so effective that there have been no fatal bites from Sydney funnel-web spiders since their antivenom was developed!

In this case, maybe it's best if we just move on before we need antivenom at all!

Where to next, Mom?

That's right! You'd better run, yanks!

Let's stop in Tanzania so you can meet a friend of mine!

WOOSH

Are we gonna meet something else that wants to eat us?

Oh, don't worry! Tarantulas are super chill.

Really? I thought tarantulas were dangerous!

Most of them just want to be left alone! They're extremely shy, introverted animals!

Hey, Peli? Are you home?

Tepi? Is that you? *OH JEEZ!* Visitors! My place is a total mess!

Kids, this is my friend *Pelinobius muticus*, the king baboon spider!

Oh gosh! Hi, I guess!

I was just about to tell them that tarantula venom isn't particularly dangerous to humans!

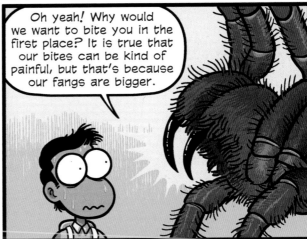

Oh yeah! Why would we want to bite you in the first place? It is true that our bites can be kind of painful, but that's because our fangs are bigger.

Old-world tarantulas are sometimes called aggressive, but a better term would be "defensive"! They will generally bite humans only when they feel threatened. Tarantulas usually prefer running away rather than trying to fight something much bigger than themselves!

Indian Ornamental Tarantula
Poecilotheria regalis
7–10 cm (2.76–3.94 in)

But when running isn't enough, new-world tarantulas have a special defense mechanism! Their abdomen is covered in a layer of special hairs that they may brush off at a predator when threatened.

Costa Rican Tiger Rump
Davus fasciatus

These hairs, called urticating hairs, can lodge themselves in the skin or eyes of an attacker, causing extreme irritation and pain. Their effectiveness varies depending on the species of tarantula.

Hey! Watch it, pal!

The Chilean rose tarantula, *Grammostola rosea*, has hairs that are only mildly irritating to humans.

Whereas the Brazilian whiteknee tarantula, *Acanthoscurria geniculata*, has hairs that can be extremely painful.

Antilles Pinktoe Tarantula
Caribena versicolor

Tarantulas may also use these hairs to mark territory. Some, like the tree-dwelling *Caribena* species, weave urticating hairs into their egg sacs to protect them against predators!

So spiders that are big and hairy are called tarantulas?

Tarantula originally referred to a specific wolf spider, *Lycosa tarantula*, named after the italian town of Taranto. It gradually became a term for any large spider!

Today it specifically means a spider who is a member of the family Theraphosidae. There are about a thousand species of tarantulas, and they're popular with people who like to keep spiders as pets!

There are a few reasons tarantulas make better pets than smaller spiders! They're much more shy and docile, and since their venom is so mild, accidental bites are less of a problem. They're mygalomorphs, as well, which means they live a lot longer!

Mexican Redknee Tarantula
Brachypelma hamorii

Speaking of big spiders, have you met *Theraphosa blondi* yet?

Ohhh, can we, Mom? I've only gotten to see them in books! *PLEASE PLEASE PLEASE?*

Hmmm, we can try! It is sort of on the way back...

Thanks, Peli! Sorry to chat and run, but there's still more to learn!

Have fun!

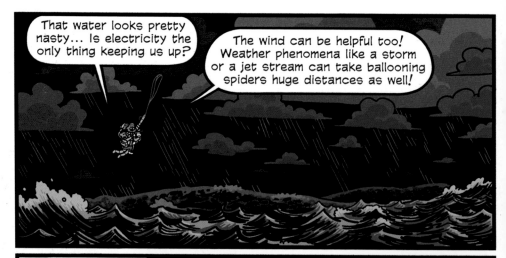

That water looks pretty nasty... Is electricity the only thing keeping us up?

The wind can be helpful too! Weather phenomena like a storm or a jet stream can take ballooning spiders huge distances as well!

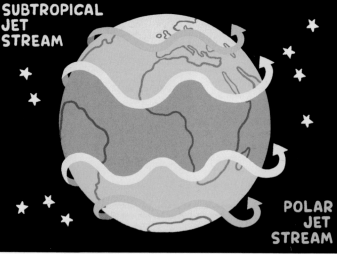

Jet streams are giant, fast-moving air currents that are forming and dissipating at all times in our atmosphere. The biggest jet streams typically form at the borders of large masses of air when the two masses are drastically different temperatures, and ballooning spiders can get caught in them to travel massive distances!

SUBTROPICAL JET STREAM

POLAR JET STREAM

Looks like we'll be reaching South America soon. Let's aim for Peru!

Awww, I thought we were going to Brazil to see the Amazon rain forest!

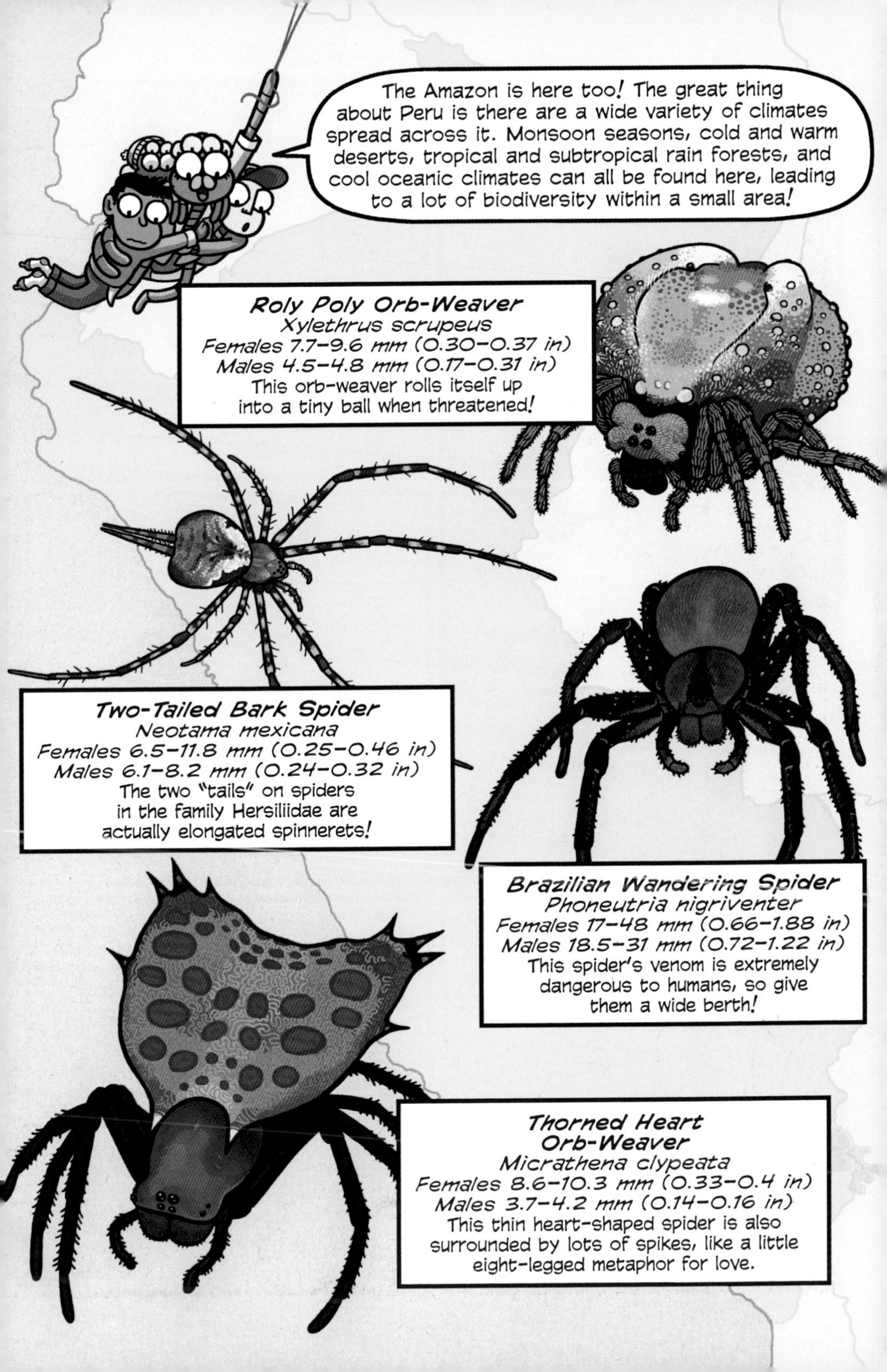
The Amazon is here too! The great thing about Peru is there are a wide variety of climates spread across it. Monsoon seasons, cold and warm deserts, tropical and subtropical rain forests, and cool oceanic climates can all be found here, leading to a lot of biodiversity within a small area!
Roly Poly Orb-Weaver
Xylethrus scrupeus
Females 7.7–9.6 mm (0.30–0.37 in)
Males 4.5–4.8 mm (0.17–0.31 in)
This orb-weaver rolls itself up into a tiny ball when threatened!
Two-Tailed Bark Spider
Neotama mexicana
Females 6.5–11.8 mm (0.25–0.46 in)
Males 6.1–8.2 mm (0.24–0.32 in)
The two "tails" on spiders in the family Hersiliidae are actually elongated spinnerets!
Brazilian Wandering Spider
Phoneutria nigriventer
Females 17–48 mm (0.66–1.88 in)
Males 18.5–31 mm (0.72–1.22 in)
This spider's venom is extremely dangerous to humans, so give them a wide berth!
Thorned Heart Orb-Weaver
Micrathena clypeata
Females 8.6–10.3 mm (0.33–0.4 in)
Males 3.7–4.2 mm (0.14–0.16 in)
This thin heart-shaped spider is also surrounded by lots of spikes, like a little eight-legged metaphor for love.

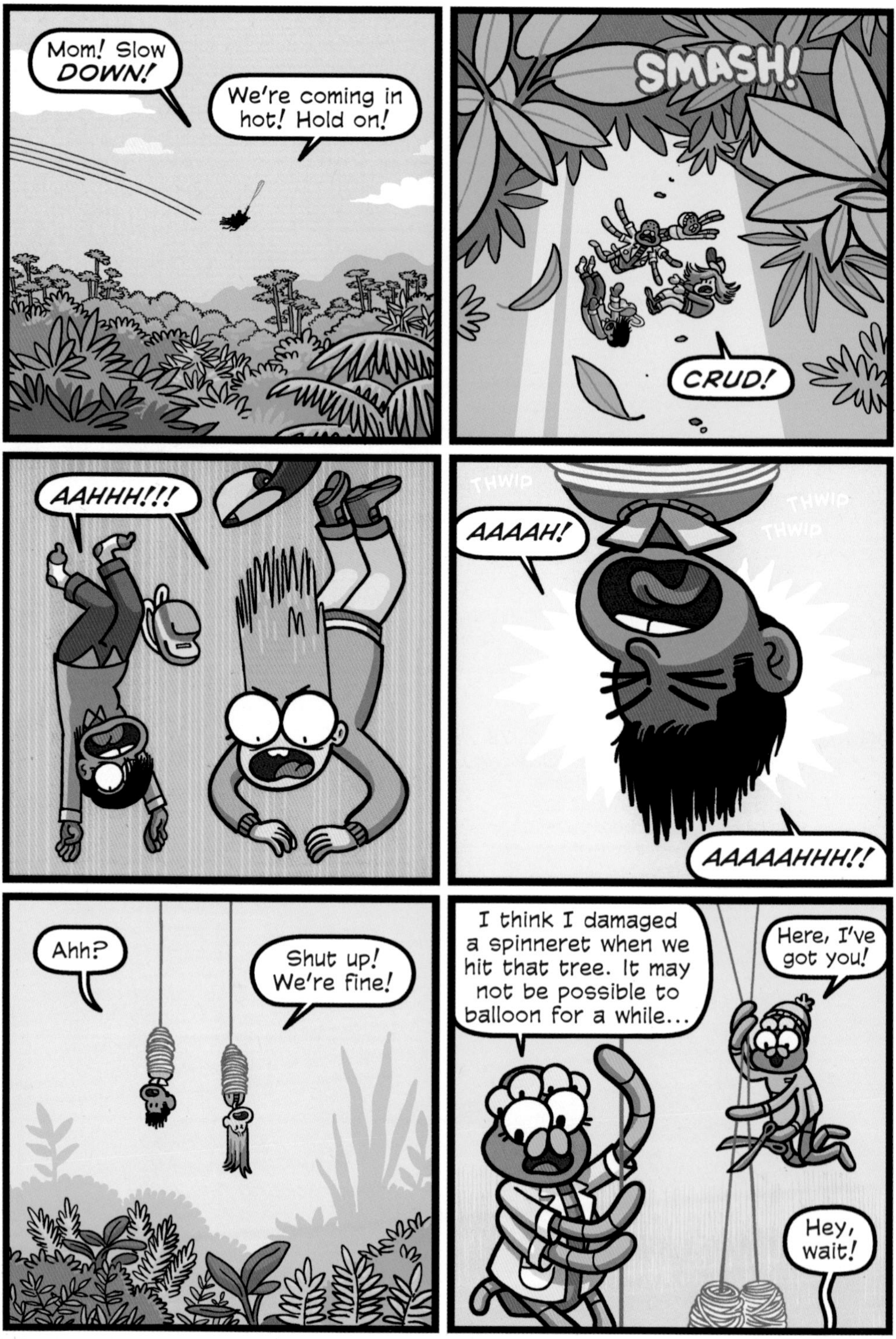
Mom! Slow DOWN!
We're coming in hot! Hold on!
SMASH!
CRUD!
AAHHH!!!
THWIP
AAAAH!
THWIP
THWIP
AAAAAHHH!!
Ahh?
Shut up! We're fine!
I think I damaged a spinneret when we hit that tree. It may not be possible to balloon for a while...
Here, I've got you!
Hey, wait!

We're close enough to home now that we might be able to walk back, but it could take a long time!

But while we're here, let's meet one last spider. The biggest in the world, in fact!

SMAK

THUD

Wait, I thought you said the biggest spider in the world was the giant huntsman!

They have the widest leg span, but this titan of spider-kind is the biggest overall!

Yeah, they are *HUGE!*

Keep an eye out, this kind of marshy land is one of their favorite places to hunt!

Does this spider have a common name too?

It sure does!

THE GOLIATH BIRDEATER!

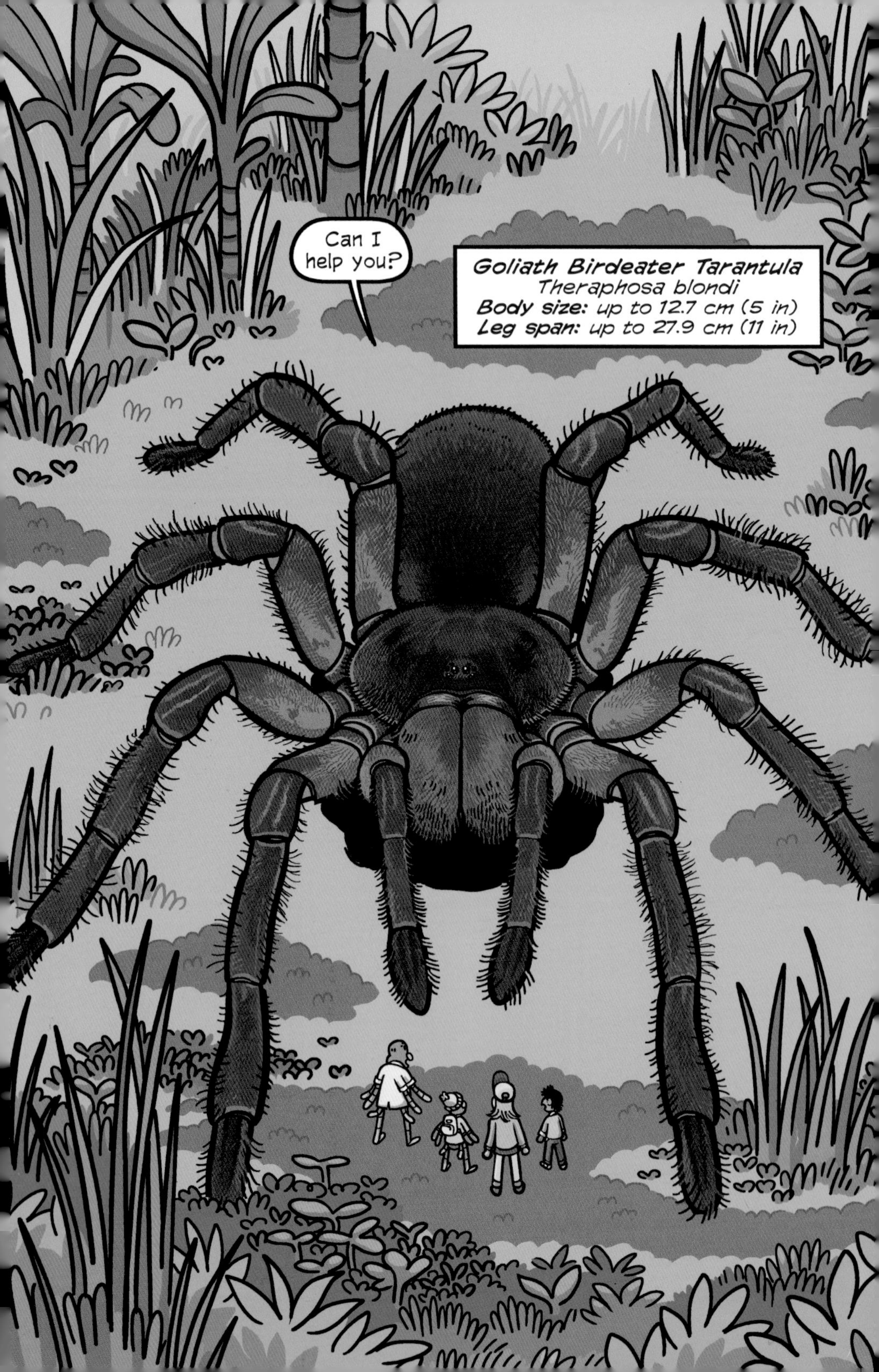
Can I help you?
Goliath Birdeater Tarantula
Theraphosa blondi
Body size: up to 12.7 cm (5 in)
Leg span: up to 27.9 cm (11 in)

≥*Gulp*≥ Uh...we didn't realize you were right behind us! We'll let you get back to eating things much smaller than you, bye!

Wait! Don't leave! I won't eat you puny little morsels!

Wait, really?

Not until I get really hungry, at least! I usually eat stuff a little bigger than you, like crickets and cockroaches, or stuff like mice, lizards, and frogs when I can get my 'palps on them!

Fascinating! This is better material for my research than I could have hoped for!

So you're called birdeater, but you don't eat birds?

Not usually! Sometimes we get lucky and catch a small one, but birds can fly, and tarantulas, especially giant ones, are way too big for ballooning!

Goliath birdeaters are big, but they aren't really harmful to humans! Their venom has been compared to a wasp sting, which would hurt, but in reality, humans are more dangerous to them!

In the last 50 years, over 20% of the Amazon has been deforested, mostly to make room for cattle for the beef and leather industries.

It continues to be clear-cut and burned at an accelerating rate, making up 14% of the annual deforestation across the globe.

Deforestation, pollution, and rising temperatures on earth, sometimes called global warming or climate change, are all leading to the destruction of spiders' natural habitats across the world.

GLOBAL WARMING

Greenhouse gases like CO_2 from coal or diesel, or methane from cow manure at industrial farms, are trapped in the atmosphere. Solar radiation that would normally bounce off the earth and leave our atmosphere is trapped by these gases and radiated back down, causing global temperatures to rise.

And massive changes for one animal means massive changes for all the animals around them. Every ecosystem on earth holds a delicate balance between every plant and animal that inhabits it!

Removing one species of spider can have a ripple effect that doesn't stop at the borders of that ecosystem! Animals that eat that spider suddenly have less food, and anything the spider eats will experience a huge population boom. Those changes then affect more plant or animal populations, and so on and so on.

Spiders collectively eat **880 MILLION TONS** of insects each year! That means they eat the weight of about 12,000 blue whales every single day!

KEEP 'EM COMIN'

And insects aren't just annoying! They can destroy crops, ruin gardens, damage homes, and more! Without spiders, humans would be overwhelmed by a never-ending swarm of bugs at all times!

Many of the insects that spiders eat are carriers of bacteria or disease as well, meaning that spiders actually help stop the spread of sickness!

Oooh, a buffet!

Mosquitoes can carry malaria, an infectious disease that can be fatal to humans.

Ticks can spread a serious infection called Lyme disease by biting.

And flies are capable of bringing a huge range of diseases to humans, like dysentery, cholera, and tuberculosis.

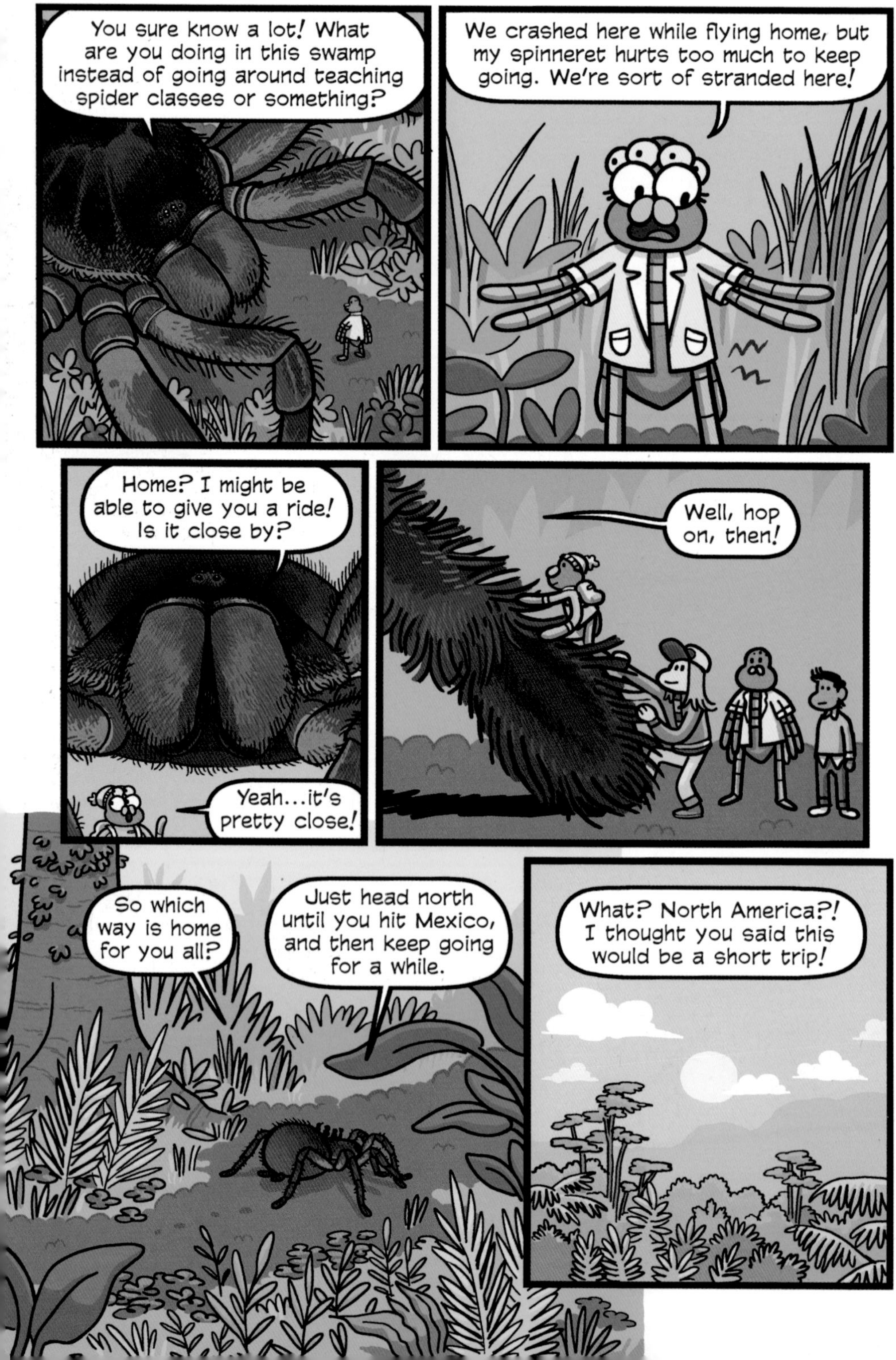
You sure know a lot! What are you doing in this swamp instead of going around teaching spider classes or something?
We crashed here while flying home, but my spinneret hurts too much to keep going. We're sort of stranded here!
Home? I might be able to give you a ride! Is it close by?
Yeah...it's pretty close!
Well, hop on, then!
So which way is home for you all?
Just head north until you hit Mexico, and then keep going for a while.
What? North America?! I thought you said this would be a short trip!

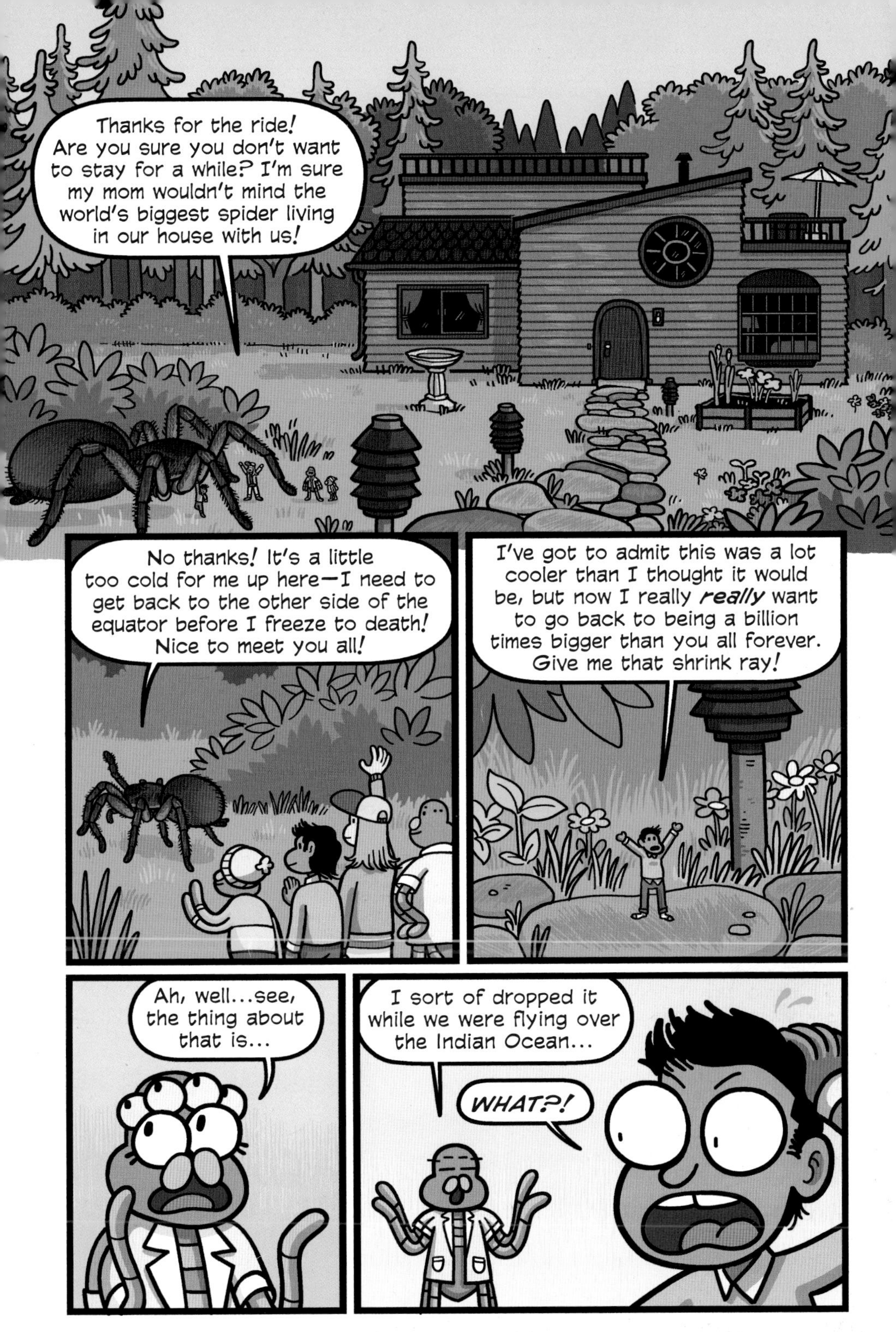
Thanks for the ride! Are you sure you don't want to stay for a while? I'm sure my mom wouldn't mind the world's biggest spider living in our house with us!
No thanks! It's a little too cold for me up here—I need to get back to the other side of the equator before I freeze to death! Nice to meet you all!
I've got to admit this was a lot cooler than I thought it would be, but now I really *really* want to go back to being a billion times bigger than you all forever. Give me that shrink ray!
Ah, well...see, the thing about that is...
I sort of dropped it while we were flying over the Indian Ocean...
WHAT?!

Just kidding!

Hahaha!

Don't joke about that!

ZAP

Yes! It's good to be an apex predator again!

I can't believe we have to say goodbye now—it feels like we met just yesterday!

That *was* yesterday! Come visit us anytime, we'll be right downstairs!

Bye! Come hang out in my room whenever you want—I've got all kinds of dead bugs in there!

Sounds... awesome...

Jeez, we really have been gone a whole day. I hope Mom isn't mad...

WHAT DO YOU THINK YOU'RE UP TO, WALKING IN THE DOOR AT SIX IN THE MORNING?! YOU ARE BOTH IN SO MUCH TROUBLE!

THE END

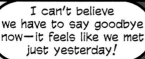

—GLOSSARY—

Abdomen
> The abdomen, or opisthosoma, is the back segment of a spider's body, which contains its heart, silk glands, and important parts of its digestive system.

Aggregate
> The sticky glue that holds prey captive when they make contact with an orb web.

Amino acids
> Organic compounds that combine to form proteins and are often referred to as building blocks of life in the natural world.

Antivenom
> May also be called antivenin or venom antiserum. A medication made to neutralize toxins in the bloodstream.

Arachnid
> A classification of joint-legged invertebrate animals known as arthropods that includes spiders, ticks, mites, and scorpions.

Arachnologist
> A person who studies arachnids.

Arachnophile
> A person who loves spiders and/or other arachnids.

Araneomorph
> A suborder of spider that is usually smaller in size and has chelicerae that move in a side-to-side pinching motion. Sometimes called Labidognatha.

Arthropod
> An invertebrate animal from the phylum Euarthropoda, which includes all insects, arachnids, and crustaceans.

Biodiversity
A term referring to the wide variety of life-forms on earth, or sometimes just the species within a specific ecosystem or area.

Calamistrum
A row of special bristles on the legs of cribellate spiders used to comb bands of fuzzy silk out of the silk-spinning organ known as the cribellum.

Cephalothorax
The front segment of a spider's body, which contains its eyes and mouth opening. Sometimes called a prosoma.

Chelicerae
The front pair of appendages on a spider's body, which contain the fangs and often venom glands.

Cribellum
A plate covered in tiny holes that extrude many thin strands of silk, which are combed out by the specialized leg bristles known as calamistrum. Cribellate spiders are any spider that has a cribellum.

Exoskeleton
An external skeleton that protects the body of an animal, as opposed to an endoskeleton, like that of a human, which resides inside the body.

Gland
An organ inside an animal that synthesizes chemicals or substances for use inside the body or to be secreted outside the body.

Hemolymph
Invertebrates have hemolymph instead of blood. Blood (which is found in vertebrates) gets its red color from the interaction between iron and oxygen inside of the hemoglobin in red blood cells. But hemolymph is light blue because it contains hemocyanin, which contains copper instead of iron.

Hydrogel
A gel made up of polymers that are usually suspended in water. They are highly absorbent and have been used to create things like contact lenses and wound dressings.

Invertebrate
  An animal without a backbone or
  any kind of spinal column.

Larva
  The first stage of life (after hatching) for an insect
  before going through its metamorphosis to become
  an adult.

Mesothelae
  The closest living relatives of the first spiders on earth belong to
  this dwindling suborder.

Mygalomorph
  A suborder of spider that is usually larger in size and has chelicerae that
  move in an up-and-down motion. Sometimes called Orthognatha.

Myrmecophily
  A term for any species of organism, including plants and fungi, that
  interact positively with ants or ant colonies.

Pedipalp
  The second pair of appendages on a spider, on either side of the
  chelicerae, which resemble shortened legs. Males have bulbs on the
  end of their pedipalps, which is an easy way to visually differentiate
  males and females.

Pheromones
  Chemicals produced by an animal that affect the behavior of other
  members of the same species.

Phototaxis
  Bodily movement by an organism in response to changes in light.

Pigment
  A material that reflects light at a specific wavelength, giving it color.
  Melanin is an amino acid that is the pigment responsible for most skin
  and hair coloration in humans.

Protein
  A complex molecule made up of amino acids, one of the main
  components of all organic tissue.

Seta

A stiff, hairlike bristle usually found on invertebrates. Setae is the plural form.

Spinneret

The organ on a spider and some insect larvae that produces silk.

Stabilimentum

A wide silk structure built on some orb webs that may reflect UV light and provides camouflage for a spider.

Tapetum

A reflective surface in the eye that helps increase available light and improves one's ability to see in dark conditions.

Toxin

A poisonous, usually unstable substance produced by a living organism or a living cell.

Trichobothria

Elongated setae found on arachnids and insects that can detect vibrations and electrical charges in the air.

Venom

A poisonous substance secreted by animals and typically injected into prey or aggressors by biting or stinging.